PLOUGHING A NEW FURROW

A BLUEPRINT FOR
WILDLIFE-FRIENDLY FARMING

Malcolm Smith

Whittles Publishing

Published by
Whittles Publishing Ltd.,
Dunbeath,
Caithness, KW6 6EG,
Scotland, UK
www.whittlespublishing.com

© 2018 Malcolm Smith
ISBN 978-1-84995-328-3

"To husband is to use with care, to keep, to save, to make last, to conserve. Old usage tells us that there is a husbandry also of the land, of the soil, of the domestic plants and animals - obviously because of the importance of these things to the household. Husbandry is the art of keeping tied all the strands in the living network that sustains us. And so it appears that most and perhaps all of industrial agriculture's manifest failures are the result of an attempt to make the land produce without husbandry."

WENDELL BERRY (AMERICAN NOVELIST, POET, ENVIRONMENTAL ACTIVIST, CULTURAL CRITIC ... AND FARMER) IN *BRINGING IT TO THE TABLE: ON FARMING AND FOOD* (COUNTERPOINT, 2009)

CONTENTS

ACKNOWLEDGEMENTS

My primary thanks go to the farmers I have met in the course of writing this book. They have been unfailingly courteous, helpful and informative; we have had constructive discussions in many a farmhouse kitchen and out on their land. And they have been very willing to show me around so that I could see the work they have been doing and the commitment they have made in improving their farms for wildlife. I have tried to reflect in these pages many of the ideas they raised with me. Some of the farmers I met will, no doubt, be unhappy with all of my conclusions but I have reached them in an attempt to find a better way in which Britain's farms can, once again, be havens for wildlife and support productive farming.

I thank John Cherry, Weston Park Farms, Stevenage, Hertfordshire; Ian Dillon, Farm Manager, RSPB Hope Farm, Cambridgeshire; Will Dracup, Broadaford Farm, Widecombe-in- the-Moor, Newton Abbot, Devon; Jake Freestone, Farm Manager, Overbury Estate, Overbury, Gloucestershire; Peter French, Deane Farm, Deane Lane, Maidencombe, Devon; Tom Harris, Ffosyficer, Abercych, Boncath, Pembrokeshire; Andrew Hattan, Low Riggs Farm, Middlesmoor, Harrogate, North Yorkshire; Stephen James, NFU Cymru Chairman, Pembrokeshire; Robert Kynaston, Great Wollaston Farm, Halfway House, Shrewsbury; John Lloyd, Cynghordy Hall, Cynghordy, Llandovery, Carmarthenshire; Tom Lord, Lower Winskill Farm, Langcliffe, Settle, North Yorkshire; Roy and Irene Newhouse, New House Farm, Malham, North Yorkshire; Don Osborne and Rupert Weatherall, Leasefield Farm, Beaworthy, Devon; Tony Reynolds, Thurlby Grange Farm, Lincolnshire and Paul Simpson, Newlands Farm, West Lulworth, Wareham, Dorset.

I received help, information, comments and suggestions from a wide range of farming experts, conservationists, ecologists, academics and scientists as well as from Governments and Government Agencies across the UK. I thank:

Shelley Abbott, Conservation Grade, St Neots, Cambridgeshire;

Dr Kevin Austin, Deputy Head Agriculture, Sustainability and Development Division and Tristan Rutherford, Head of Glastir Advanced, Rural Payments Wales, Welsh Government;

Sallyann Baldry, Business Development Director and Jane Cummins, Skills View Newsletter Editor, Lantra;

Simon Bareham, Principal Adviser, Business, Regulation & Economics, Natural Resources Wales;

John Bates, Senior Media & PR Manager, Levy Payer Communications, Communications & Market Development, Agriculture & Horticulture Development Board;

Jeni Beaumont, Farming Advice Service;

Tim Blackstock, former Head of Habitats and Species, Countryside Council for Wales;

Catherine Brown and Katherine MacMillan, Communications Greener and Mark Taylor, Senior Media Manager, Communications Healthier, Scottish Government;

Ariel Brunner, Senior Head of Policy for Europe and Central Asia, BirdLife International;

Tara Challoner, Upper Nidderdale Wildlife and Farming Project Officer, Yorkshire Wildlife Trust and Iain Mann, Scheme Manager, Upper Nidderdale Landscape Partnership;

John and Pamela Clarke, Kemerton, Worcestershire;

Wayne Coles, Media Manager, Customer Directorate, Rural Payments Agency, Reading;

Professor Christl Donnelly, Department of Infectious Disease Epidemiology, Imperial College London;

Simon Draper, Farming Advice Service;

Caroline Drummond, CEO, Linking Environment and Farming (LEAF);

Professor Matthew Evans, Professor of Ecology, School of Biological and Chemical Sciences, Queen Mary University of London;

Dr Jenny Gibbons, Senior Dairy Scientist and Ray Keatinge, Head of Animal Science, Agriculture & Horticulture Development Board;

Professor Dave Goulson, School of Life Sciences, University of Sussex;

David Harvey, Emeritus Professor of Agricultural Economics & Editor, Journal of Agricultural Economics, Newcastle University;

Professor David Hopkins, Dean of Agriculture, Food & Environment, The Royal Agricultural University, Cirencester;

Peter Howells, Farm Policy Advisor and Sarah Jones, Communications Adviser, NFU Cymru;

Steven Jacobs, Business Development Manager, Organic Farmers and Growers;

Laurie Jackson, Farm Pollinator and Wildlife Advisor, Buglife;

Dr Robin Jackson, Senior Manager (Land Based Industries), Portfolio Management & Development, City and Guilds;

Cath Jeffs, Cirl Bunting Project Manager, RSPB Exeter, Devon;

Amir Kassam, Visiting Professor in the School of Agriculture, Policy and Development, University of Reading;

Joseph Keating, Knowledge Exchange Manager Beef & Lamb, Technical Directorate, Agriculture & Horticulture Development Board;

Clunie Keenleyside, Senior Fellow, Institute of European Environmental Policy;

Tom Lancaster, Senior Land Use Policy Officer and Tony Morris, Senior Scientist, RSPB;

Matt Lobley, Associate Professor in Rural Resource Management and Director Land, Environment, Economics and Policy Institute (LEEP), University of Exeter;

Dr Lois Mansfield, Principal Lecturer, The University of Cumbria;

Alan Matthews, Professor Emeritus of European Agricultural Policy, Department of Economics, School of Social Sciences and Philosophy, Trinity College Dublin;

Thomas McCabe, Project Manager and Ryan Pike, Statistician, TB Team, Office of the Chief Veterinary Officer, Welsh Government;

Davy McCracken, Professor of Agricultural Ecology and Head of Hill & Mountain Research, Scotland's Rural College;

Chloe McGregor, Environment Manager, Dairy UK;

Peter Melchett, Policy Director, Natasha Collins-Daniel, Press Manager, Emily McCoy and Hayley Coristine, Digital Communications & Press Officers, The Soil Association;

Chris Moody OBE, Chief Executive, Landex (Land based Colleges Aspiring to Excellence), University of Northampton;

Jean Murphy, British Wool Marketing Board;

Emily Newton, Conservation Officer, Dorset Wildlife Trust, Kingcombe, Dorchester, Dorset;

Dr. Joe Nunez-Mino, Director of Communications & Fundraising, Bat Conservation Trust;

Tom Oliver, Associate Professor Landscape Ecology University of Reading;

Karen Orr, Environmental Farming Branch, Herbie Jones, Head of Countryside Management Delivery Branch and Dr Richard Crowe, Department of Agriculture, Environment and Rural Affairs, Northern Ireland;

Adriana De Palma, PREDICTS Postdoctoral Research Assistant, Natural History Museum, London;

Brian Pawson, Senior Agriculture Advisor, Natural Resources Wales;

Richard Percy, Chairman NFU Mutual, Cottingham Farm, Flaunden Lane, Hemel Hempstead;

Jenny Phelps, Senior Farm Conservation Advisor, Gloucestershire Farming & Wildlife Advisory Group;

Anthony Pope, Conservation Agriculture Consultant;

Jane Salter, The Agricultural Industries Confederation;

Lisa Schneidau, Northern Devon Nature Improvement Area Project Manager, Devon Wildlife Trust, Beaworthy, Devon;

Cath Shellswell, Farmland Adviser and Save Our Magnificent Meadows Adviser at Plantlife;

Kathryn Smith, Project Manager, FarmWildlife;

Alan Spedding, Editor, *Farming Futures*;

Mike Thomas, Media Adviser, NFU Communications;

Dr Maria Tsiafouli, Department of Ecology, School of Biology, Aristotle University, Thessaloniki, Greece;

Dr Martin Warren, Chief Executive, Dr Sam Ellis, Director of Conservation and Regions and Jenny Plackett, Southwest Senior Regional Officer, Butterfly Conservation;

Ben Watts, Partner, Kite Consulting;

Dr Nicola Whitehouse, Associate Professor in Physical Geography, Plymouth University;

Ian Wilkinson, Managing Director and Fiona Mountain, PR & Communications, Cotswold Seeds Ltd;

Dr Ben Woodcock, Ecological Entomologist, Centre for Ecology and Hydrology;

Alison Wray, Analyst, Farming Statistics, Food and Farming Directorate, Defra.

I also want to thank Keith Whittles for his confidence in this project, for his advice and guidance on various aspects as it developed from concept to completion and Rachel Oliver, my editor at Whittles, who sharpened up my chapters and gave me lots of constructive comments. If there are errors, as there surely will be, they are my responsibility entirely.

GLOSSARY

With just 0.74% of the UK's population engaged in farming – and most people's experience of it limited to buying a pre-packed piece of meat, a plastic container of milk or a packet of well-washed carrots in a supermarket – the dichotomy between those who produce much of our food and everyone else in society has never been greater.

Consequently I thought it would be helpful to define many of the terms used in this book; terms often used routinely by farmers and farming policy experts but which many people will be less familiar with. Some are technical; others are perhaps self evident but I have included them for completion:

agri-environment climate schemes: Agreements negotiated voluntarily by farmers with the UK's agriculture departments in which a set of wildlife protection measures and other land management modifications to protect water, historic features and public access are agreed in return for an annual payment.

anthelmintic: A drug used to destroy parasitic worms in livestock.

arable: Land capable of being ploughed and used to grow crops.

autumn sown crop: A crop sown in autumn and left growing over winter to be harvested the following summer. No stubble (low cut stems) is left in the ground after harvest; instead, stubble is ploughed into the soil .

Basic Payment Scheme (BPS): The direct payments – subsidies – paid to farmers to keep farming their land funded from Pillar 1 of the Common Agricultural Policy. It replaced the Single Farm Payment Scheme after 2013.

Birds Directive: European Union Directive 2009/147/EC which makes it mandatory for all member states to protect and maintain their populations of all wild bird species across their natural range.

black grass: *Alopecurus myosuroides* is an annual grass, found on cultivated and waste land, often growing in tufts up to 80cm high. A major weed of cereal crops, it produces a large amount of seed which is shed before the crop is cut and it has developed resistance to a range of herbicides used to control it. It can seriously reduce crop yields.

blanket bog: areas of often extensive upland with a climate of high rainfall allowing peat to develop not only in wet hollows but over large expanses of undulating ground often dominated by vegetation rich in Sphagnum mosses, cotton grasses and heaths.

bovine tuberculosis: A contagious disease caused by a bacterium that can infect many animals including cattle. It is the main reason for the pasteurisation of milk.

Cereal: A cultivated crop grown for its grain such as wheat, barley, oats and maize.

CAP: Common Agricultural Policy.

catch crop: A fast-growing crop that is grown between successive plantings of a main crop to make more efficient use of growing space. Often grown over a short season, they can also help prevent runoff from bare soils.

common grazing: Often large areas of upland, moor or heathland where sheep (sometimes cattle) from several different farms graze freely without fences or walls.

Conservation Grade: An organisation that accredits food brands and other farm products that use 'Fair to Nature' farming practices to boost farm wildlife. Accredited farms have to create specific wildlife habitats on at least 10% of their land and follow a sustainability protocol.

conservation headland: Headlands of cereal crops that are left unsprayed with herbicides to allow small populations of wild plants and their associated insects to develop.

Continuing Professional Development (CPD): The process of tracking and documenting the skills, knowledge and experience gained formally and informally as people work beyond any initial training: a record of what you experience, learn and then apply.

coppice: Shoots of tree re-growth arising from stumps of trees which are cut on a regular cycle.

Countryside Stewardship: England's agri-environment climate scheme running from 2015. It replaced Environmental Stewardship.

cross-compliance: In order to receive BPS direct payments, farmers are required to follow rules on food safety, animal health, plant health, the climate, the environment, the protection of water resources, animal welfare and the condition in which farmland is maintained.

culm grassland: Wet, often waterlogged grasslands on poor draining, often clay-based soils dominated by purple moor grass and rich in a wide variety of plants and invertebrates. Now confined to NW Devon, South Wales and SW Scotland, only a tiny fraction of its original extent survives. Some has extensive willow and birch scrub.

dairy cow: Cow of a breed for milk production as distinct from beef or dual purpose breeds.

DEFRA: Department for Environment, Food and Rural Affairs responsible for agriculture and environment including wildlife conservation in England. Wales, Scotland and Northern Ireland have their own equivalent Department or Departments.

direct drilling: A method of introducing seed into the soil through the stubble of the previous crop and without the need to plough the soil first.

disc harrow: a farm implement pulled by tractor that is used to shallow till the soil where crops are to be planted. Also used to chop up unwanted weeds or crop stubbles.

downland: Gently rolling open countryside dominated by long established grasslands.

Ecological Focus Area: Every farmer claiming BPS and with more than 15 hectares of arable land is obliged to have 5% of his arable land covered by ecological focus areas such as buffer strips, forested areas, fallow land and areas with nitrogen-fixing crops.

fallow: Land that is being given a temporary rest from crop production, often for a year.

Farming and Wildlife Advisory Group (FWAG): A charitable organisation established by farmers in the 1960s to provide trusted, independent environmental advice to the farming community.

ffridd: see Intake

genetically modified crops/GMOs: A crop (or other organism) in which GM technology has enabled plant breeders to bring together in one plant useful genes from a wide range of living sources, not just from within the crop species or from closely related plants which traditional plant breeding has long done.

Glastir: Agri-environment climate scheme that started in Wales in 2014.

greenhouse gas (GHG): Greenhouse gases such as water vapour, methane and carbon dioxide stop heat escaping from the Earth into the atmosphere, leading to global warming. Nitrous oxide and methane are the main GHGs from farming.

greening: A payment additional to BPS for three sets of measures that must be adopted by all farmers claiming the basic payment: ecological focus areas; crop diversification and permanent grassland.

Habitats Directive: European Union Directive 92/43/EEC which makes it mandatory for all member states to conserve a wide range of rare, threatened or endemic animal and plant species and their habitats. Some 200 rare and characteristic habitat types are targeted for conservation.

hay: Dried grass and other plants used for animal feed. It is cut, left to dry in the field and then baled. It is fed to livestock through the winter when fresh grass is not available. Nowadays rarely gathered except for horses because its production is unreliable in the often damp UK climate. It has been very largely replaced with silage.

haylage: Silage made from grass that has been partly dried, in effect half way between silage and hay.

headland: The area at each end of a planted field of crops used for equipment turning (but see also conservation headland).

hefting: The acclimatising of a flock of hill sheep to 'their' part of hill land. A hefted flock is worth more to a farmer than one that has not been acclimatised because the animals roam far less and are easier to manage.

heifer: A young female cow.

herbal ley: A planted agricultural grassland that contains a diverse range of grasses, herbs and clovers. It produces well-balanced forage and not just large volumes of grass. Many of the species used are deep-rooting and have the ability to unlock resources from deep in the soil. Herbal leys can tolerate droughts and do not need much fertiliser input.

herbicides: Chemicals (both natural and synthetic) used to control or destroy weeds.

improved grassland: Grassland that has had some agricultural improvements made to it through drainage, the use of fertilisers and selective herbicides to increase its productivity or by ploughing and/or reseeding it with a cultivated ryegrass and/or clover. It is of extremely limited wildlife value.

inbye: The most productive land on an upland farm, mainly walled or fenced pastures, meadows and sometimes a few arable fields. It is closest to the farm and used mainly for livestock grazing or cut for silage, haylage or hay.

intake: Land on an upland farm which lies between the inbye and open fell, mountain or moor. Intake is made up of pieces of common or other land which has been enclosed from the open fell, literally taken in. It might consist of rough pasture, some bracken and areas of scrubby woodland. Called ffridd in Wales; outbye in some parts of Britain.

intensive farming: Producing the maximum number of crops in a year with a high yield from the land available and/or maintaining a high stocking rate of livestock.

LEAF (Linking Environment and Farming): An organisation promoting sustainable agriculture, food and farming helping farmers produce good food with care and to high environmental standards identified in-store by the LEAF Marque logo.

legume: A plant species that can fix nitrogen by a symbiotic relationship with bacteria that live in its root nodules. Included are clovers, trefoils, peas, beans, vetches and lucerne (alfalfa).

ley: Short-term agricultural grassland, usually sown as part of an arable rotation to provide hay, silage and grazing for a few years. Most will consist of ryegrass and clover and require significant fertiliser inputs. It is of very limited wildlife value (see herbal ley).

meadow: Grassland that is kept specifically to be cut for hay or silage although the aftermath – the re-growth of grass after the hay is cut – might be grazed outside summer. It is often rich in wild plants and invertebrates.

neonicotinoids (NNIs): Insecticides often used as crop seed coatings (and some as foliage sprays) which are absorbed into all parts of the developing plant and are implicated in the decline of some pollinating insects.

Nitrate Vulnerable Zone (NVZ): Areas of land designated as being at risk from agricultural nitrate pollution from fertilisers or manures.

no till: Growing crops or pasture from year to year without disturbing the soil through tillage, i.e. by ploughing the soil.

NGO: Non-governmental organisation, a not-for-profit organization that is independent from governments, usually funded by donations and its membership.

oil seed rape: A bright yellow-flowering crop cultivated mainly for its oil-rich seed, the third-largest source of vegetable oil in the world.

open field system: Large open fields, usually three to a village, and divided into regularly ploughed strips of barley, wheat, rye, pulses or pasture. This was common in Saxon times.

organic farming: A system of farm management and food production conforming to high standards that uses no synthetic pesticides and fertilisers and combines best environmental practices, the preservation of natural resources, wildlife conservation and the application of high animal welfare standards.

outbye: see **intake**

pasture: Grassland that is principally grazed rather than being cut for hay or silage.

permanent pasture: Pasture that has been, or is intended to be, managed for many years without being ploughed up and reseeded.

pesticides: Chemicals used to control or destroy crop pests. They include insecticides, herbicides (aimed at weeds), molluscicides (aimed at slugs and snails), and fungicides (aimed at fungi).

re-introduction: The returning of a species to a habitat or location in which it formally occurred naturally but where it became extinct.

rough grazing: Grazing of unmanaged grassland, heather and other vegetation growing on mountain slopes and moorland.

ruminant: Mammals such as cows and sheep that obtain nutrients from plant food by fermenting it in a specialized stomach prior to digestion, principally through microbial action. The process typically requires the fermented material to be regurgitated and re-chewed.

semi-natural grassland: Grassland where cutting, grazing or burning prevents scrub or trees becoming established but is otherwise unaltered by, for instance, drainage or fertiliser use. It has a greater range of plant species and can be of significant wildlife value.

sett: A badger breeding den which usually consists of a network of underground tunnels and numerous entrances, the largest spacious enough to accommodate 15 or more animals. They are often located in woodland.

silage: Grass or other crops that have been cut, allowed to wilt but not completely dry out, and are then preserved in plastic wrapping or in a large mound or pit (called a clamp) from which all air is excluded. Silage is fed to livestock through the winter when fresh grass is not available.

Site of Special Scientific Interest (Area of SSI in Northern Ireland): A UK protected area providing legal protection for habitats and species of particular note.

spring sown crop: A crop, usually a cereal, sown in the spring, harvested in summer and its stubble (the low cut stems) left undisturbed in the soil over winter before ploughing the following spring.

store cattle: Animals for beef which have been reared on one or more farms, and then are sold to dealers or other farmers who feed them up for eventual slaughter.

stubble: The short stalks from a harvested crop left in the ground after harvesting.

tillage: The preparation of soil by mechanical agitation of various types such as digging, ploughing and harrowing.

weed: A plant considered undesirable in a certain location: a plant in the wrong place.

wildwood: The wholly natural wooded landscape which developed across major parts of Britain after the last ice age unaffected by human intervention. It no longer exists as such.

CHAPTER 1

INTRODUCTION

I'm standing by the side of a small country road in mid Wales, looking out over a set of grassy pastures. It's a spot that I have been coming to for over half a century. There is nothing very distinctive about the fields either side of me; they are grazed most of the year by beef cattle and sheep.

When I first knew them, the pastures on one side of the road – drier because they are naturally well drained – were frequently visited in winter by large flocks of feeding Lapwings accompanied without fail by Starlings weaving in and out between their much larger companions. Of more interest was an occasional feeding group of Golden Plover, attractive birds that even in their winter plumage retain the distinct golden brown tinge that their name suggests.

The pastures on the opposite side of the road were not nearly as well drained; in consequence, large parts were rush-covered and boggy. In spring, the air above them was alive with Lapwings calling incessantly during their acrobatic flight displays. Here in early summer I first watched an adult Lapwing feigning a wing injury and hobbling forlornly to take my attention away from its young brood. Snipe sometimes displayed in the sky above, their ducking and diving flight causing the humming throb of their tail feathers to wax and wane curiously. Skylarks practiced their chorus unseen in the bright summer skies, and I'm sure many foxes and other mammals scoured the land in the dead of night.

Sometime in the 1980s, drainage pipes were installed under the wetter pastures, which were then ploughed and seeded with a rye grass/clover mix. The rush-filled pastures lost their sheen of pale browns and olive to become a uniform paint-pot green. The bulk of the cost of the scheme was almost certainly EU funded. So nowadays, both these pastures and the drier ones across the road look much the same, although in recent years some rushes are getting a renewed foothold, presumably because the drainage is blocking. All of this grassland is a much brighter green than it was when I knew it first, the result of synthetic fertilisers and year upon year of livestock droppings adding soil fertility.

But the biggest change is to the wildlife. I go there now and strain to hear a summertime Skylark. There are certainly no breeding Snipe or Lapwings. There's hardly any vegetation that gets above billiard-table height. Every time I visit in winter I don't spot any flocks of feeding Lapwings and definitely not any Golden Plover. The farmer who owns these fields today – and has done for many years – tells me that he thought the breeding Lapwings would simply breed elsewhere when he put the drainage in. But 'elsewhere' has been drained and ploughed up too, so gradually the Lapwings have fewer and fewer places to breed.

While these green pastures would convince most passers-by that all is well and seemingly unchanged in our farmed countryside nothing is further from the truth. As in much of the lowlands of Britain, the wild creatures that once inhabited our farmed land have been spirited away; bit by bit, field by field, farm by farm, their living quarters have been sacrificed on the altar of increasing farm productivity.

As farming practices in our lowlands have become more and more intensive and productive, field size has increased, high levels of chemical inputs including fertilisers and pesticides have boosted crop yields, and there has been a substantial move to specialisation, with farms concentrating on livestock or crop production. Farmland wildlife has been draining away. Most farmland birds are declining; many invertebrates – including bees and butterflies – are going the same way. Almost all of our lowland flower-rich hay meadows have been ploughed up or otherwise destroyed. The reality hardly supports the frequent assertion made by the National Farmers Union that 'our farmers are the custodians of the countryside'!

Yet why do we want farmland to support wildlife? Should all our farmed land be allowed simply to produce food unfettered by any constraints imposed by the need to retain wildlife? That might at least be a debatable policy wherever other sizeable land areas can be set aside to protect swathes of native habitat and conserve wildlife, thereby leaving agriculture to get on with the rest. About 45% of the land in the US is in agricultural use; in France it is 53%, but in Britain the figure is 71%.[i] Add in urban areas, industrial land, and forestry plantations, and the UK doesn't have much space left for the so-called 'land sparing' philosophy implemented in the US where strictly protected reserves and national parks contrast with heavily exploited farmland. Instead, if we want our wildlife to thrive, we have to ensure that our farmed countryside, both lowland and upland, nurtures it far better than it has over the last half century.

In recent decades there has been increasing recognition that wildlife conservation isn't only an end in itself; it contributes to a range of so-called 'ecosystem services', to use the jargon that policy makers seem to favour. For instance, wildlife provides such essentials as crop pollination, biological pest control, good soil health, and the natural cycling of nutrients. All free services!

Farmers have the same choice as gardeners. You can be the sort of gardener who manicures every square metre; sprays his lawns regularly with selective herbicides to leave nothing but perfectly coiffed grass; removes every twig and branch that falls from garden trees; and cuts down any longer vegetation or nettle-filled corner. Or you can be the sort of gardener who mows most of his lawn; rarely or never uses herbicides on his turf but leaves a part of it to grow up, meadow-like and resplendent with some common flowers; keeps a pile of decaying logs in a shady spot; creates a shallow pond to encourage amphibians and aquatic insects; and retains a few well-grown grass and flower corners too.

Farmers can also follow the intensification route: in the uplands grazing every bit of land with sheep and cattle, even the remnants of broadleaved woodland, guaranteeing that the trees will not regenerate long term to perpetuate the wooded cover it currently provides. In the lowlands they can grow 'wall to wall' barley, oilseed rape, or maize on soils fed with fertiliser, autumn sown to leave no overwinter stubbles for birds to feed amongst; spray the growing crops with pesticides; and obliterate most hedges and other wildlife features in order to achieve the maximum acreage and production. On the other hand, they can ease back on sheep and cattle numbers grazing our uplands, leaving the less accessible areas of land to develop more rank vegetation and scrub. In the productive lowlands they can re-instate some hay cutting on permanent pasture, encouraging a cornucopia of flowers to return; plant up new hedges and keep older ones in good condition; and sow annual flower seed mixes along the margins of arable crops to create habitat for pollinating insects.

But there are two major differences between gardeners and farmers. Firstly, farmers have to make a living from their land, or at least part of their living depending on the extent to which they have diversified into other income generators such as farm educational visits or running a B&B. Secondly and uniquely, farmers are heavily subsidised by the taxpayer. The bulk of the subsidy is paid to them simply to remain as farmers; it is not means tested. There are very few restrictions imposed in order to receive it, although some farmers would argue the opposite and believe that they should be unfettered recipients of large amounts of public money. The very largest farms and farming estates receive annual subsidy payments costing millions of pounds, merely because they are farmers and food producers. Whether these subsidies will continue after Brexit and into the long term, or to what extent they might be reduced as they compete directly and very publicly with other government and public priorities, remains to be seen. For the first time, each of the four governments in the UK (agriculture is a devolved responsibility) will have to weigh farming need against the needs of the NHS, education, social support, housing, and much more besides.

If subsidies do persist in some form or other, surely the taxpayer should expect to have a say in how farms are managed and what sort of countryside is the end result? Is it not reasonable that UK taxpayers should have guarantees that the food produced is wholesome and healthy, that high animal welfare standards are the norm, and that farming practices also cater for wildlife adequately on every farm receiving any such support? After all, an estimated six million people in the UK belong to conservation organisations such as The Wildlife Trusts, the RSPB, the National Trust, the World Wildlife Fund, and others, far outnumbering the 466,000 employed in farming (though a good number of farmers belong to conservation organisations too).

For this book I have spent a substantial amount of time with a number of farmers who are determined to better balance their primary job of producing wholesome food with caring for wildlife on their farms. Our discussions, in farm kitchens and out in their fields or on windswept moors, have convinced me that they feel a genuine concern and a responsibility to nurture their land, to retain much of the wildlife they still have, and to put back a good amount of what has been lost through decades of generous EU Common Agricultural Policy (CAP) funding. Most are doing so with financial help from a part of the CAP aimed at restoring and improving farmland wildlife habitats, ironically after years of the CAP funding their destruction. They form the case examples of several chapters and I have incorporated many of their ideas and suggestions in the pages of this book.

There is, though, no virtue in glossing over the fact that, over millennia, farming has changed much of our countryside beyond all recognition. The large tracts of open moor and mountain so admired by hillwalkers, and the patchwork of fields and hedgerows in the lowlands, are both human constructs which would not appear as they do today if farming wasn't our major land use. These changes have altered enormously the habitats and species of wildlife our countryside supports; the intensification of lowland agriculture during and since World War II in particular has caused further and lasting declines in some of our most valued habitats, plants, and animals. Understanding how farming has developed and changed over millennia – and how this has impacted on our wildlife – is therefore the subject of the first two chapters of this book.

There are issues that polarise opinion: the horrendous problem of TB in cattle and whether culling large numbers of badgers will reduce its incidence; and whether pesticides such as neonicotinoids are depleting our bees and other pollinating insects. The latter is a microcosm of a larger debate about the array of pesticides being used in modern farming and the rate at which crop pests are building up resistance to some of them. Both are discussed here, and my conclusions will most certainly not please everyone on either side of these difficult debates.

The seeds were sown by the Conservative-Liberal Democrat Government early in 2013 for a referendum on whether or not the UK should remain in the EU, but the result on 23 June 2016 surprised many. The Common Agricultural Policy – although it spends a little under 40% of the whole EU budget (and has spent 70% of it in the past) – hardly featured in any debates except, presumably, amongst farmers. Seemingly, a majority of farmers voted to leave a system that has supported and subsidised them for decades and, instead, risk their chances competing for any taxpayer funding with other vital calls on Exchequer resources. Out of the 577 farmers who completed a *Farmer's Weekly* survey two months before the referendum, 58% said they wanted to leave, 31% said they wanted to remain, and 11% were undecided.[ii] Given that most profit on many upland and small lowland UK farms can be attributed to EU farm supports and that more than 90% of farm produce is exported tariff free to other EU countries, an arrangement likely to be curtailed when we withdraw, is it any wonder that some commentators have likened their decision to 'turkeys voting for Christmas'?!

When I started writing this book, it seemed unlikely that the UK would leave the EU and very probable that farmers would want to keep their CAP in place. But the CAP has been forsaken. Brexit means that by maybe 2021, the four governments in the UK will have to put in place whatever financial supports and policies for farming they consider necessary within the context of other competing priorities. No longer will conservation interests be pressing for modifications to the CAP to re-balance farm food production with wildlife conservation. Nevertheless, the influence of the CAP has been so dominant for so long, it is impossible to understand how modern farming has developed, and what wildlife damage it has caused, without my including a chapter about the policies that have underpinned and funded it.

With a CAP-free future just ahead, what's needed is a fresh look at policies and supports to achieve a far better farming/wildlife balance than has ever been achieved with the CAP in place. There is, of course, no reason to suppose that all four governments in the UK will adopt new or similar farming policies and funding supports once we have left the EU; that is for those governments to decide individually.

So Brexit provides an unprecedented and unpredictable opportunity to look anew at what the four nations of the UK – both their publics and their governments – want from our farmers. If the British public want to have a farmed environment which produces food *and* caters sensitively for wildlife then we are on the threshold of an unparalleled opportunity. By no means all farmers and agricultural commentators are going to agree with what I suggest needs to be done to achieve that. Nonetheless, the ball has started to roll; where it ends up is hard to determine. One thing is for sure: farming is set to change once more, and I very much hope that those changes mean that it caters much better in future for wildlife than it has for decades.

Endnotes
i   https://www.cia.gov/library/publications/the-world-factbook/fields/.
ii   *Farmer's Weekly*, 29 April 2016.

CHAPTER 2

FROM AXES TO PESTICIDES

*Why do farmers farm, given their many frustrations and difficulties normal to
farming? They must do it for love. Farmers farm for the love of farming.
They love to watch and nurture the growth of plants. They love to
live in the presence of animals. They love to work outdoors. They
love the weather, maybe even when it is making them miserable.*

WENDELL BERRY, AMERICAN NOVELIST, POET, ENVIRONMENTAL ACTIVIST,
CULTURAL CRITIC, AND FARMER, *BRINGING IT TO THE TABLE: ON FARMING AND FOOD*
(COUNTERPOINT, 2009)

Sometime before 6000 BC, the first farmers to ever arrive on our shores trudged
across the flat, marshy land that then linked the east coast of Britain to the European
continent. They brought with them cattle bred originally from Aurochs that they
had captured in the extensive forests of continental Europe, as well as sheep and
goats they had domesticated, neither of which were native to Britain. These early
farmers almost certainly also brought with them varieties of wheat and barley that
they had bred carefully by selection over thousands of years. Recent research – such
as that by archaeologist Dr Melinda Zeder – shows that early domestication of both
plants and animals was happening at least three millennia earlier in the Middle East.[i]

It must have been a daunting experience for the immigrant farmers. The land
was clothed in broadleaved woodland, dense in places with maybe a scattering
of more open glades. What wasn't wooded was marsh, coastal saltings, or the
highest reaches of rocky mountains above the treeline. Farmers, though, are
nothing if not tenacious.

They began by burning sections of forest, cutting trees down with their stone
axes or girdling them, chopping out a complete ring of bark around the trunk to

kill them so that the dead tree stools eventually rotted away. In all probability the small communities they established were not always permanent. Many exploited a 'slash and burn' style of shifting agriculture still commonplace in developing parts of the world today. Land was cleared and farmed until fertility was exhausted – maybe 20 years – at which point the farmers moved to clear new land and start afresh. Abandoned land would perhaps have reverted to scrub and woodland before it was cleared again.

Britain hadn't always been wooded. Around 8000 BC, the whole of the country – and much of Northern Europe – was clothed in tundra: low growing shrubs, sedges, mosses, and lichens. There were no trees. The great glaciers and snowfalls of the last Ice Age were easing, albeit slowly. As the climate slowly warmed, the first trees took root, and within 2000 years around two-thirds of Britain was clothed in forest.

According to the respected historical ecologist, the late Oliver Rackham (1939–2015), birch was the first to arrive[ii]; its seeds would have been blown in from the European continent. Always a coloniser, birch was soon replaced mainly by Scots Pines of the sort that still survive in pockets in Central Scotland. These are not to be confused with the serried ranks of conifers (mostly species imported from North America) planted mostly in our uplands by foresters over the last century or two. Hazel, elm, oaks, and alder, followed by other trees, shrubs, flowers, and grasses of the forest floor slowly established themselves. The 'wildwood' that grew up was an entirely natural forest that would have supported a welter of wildlife; it was said that a squirrel (presumably a Red Squirrel in those pre-Grey days) could cross from the coast on one side of England to the other without touching the ground!

Most forest experts used to believe that the bulk of this 'wildwood' was dark, dense, and impenetrable; a view based perhaps more on myth and hearsay than scientific fact. Certainly our earliest British ancestors didn't appreciate the forest in the way we do today. It was an untamed place, harbouring witches, fays, monsters, and magic springs; a place of mystery and danger, of things that are not what they seem. In the forest you hear rustlings, like whispers and stealthy footsteps, and it is all too easy to become lost. It was a place to be avoided. Better to stay in the open, where a man's eyes could give warning of danger.

Recent research by Dr Nicola Whitehouse and Dr David Smith suggests that by no means all of Britain's wildwood was so dense and forbidding.[iii] Some of it was probably more open, much like the New Forest in the south of England today, akin to pasture with woodland rather than dense forest. Naturally occurring fires, sparked by hot summer weather or lightning strikes, would have created temporary open areas; so too, would browsing by forest mammals such as elk and deer, which would have prevented sapling trees from growing up.

Their research relied on beetles; more precisely on the fossilised remains of the hard exoskeletons of these creatures: 'Around 6000 BC, both forest and tree beetles became more abundant while species of more open areas declined. The forest canopy was closing,' says Nicola Whitehouse. 'We don't have much evidence but I think that the forest might have been less prone to fires because the climate became wetter and also because the forest animals were hunted more regularly so grazing was reduced.

'By 4000 BC it was all change; dung beetles, species reliant on the excrement of grazing animals, became more abundant and forest beetles declined.' Early farmers were already starting to change the landscape and, with it, Britain's wildlife.

As they cleared more land for growing crops and keeping livestock, pigs were domesticated from wild boar already living in the slowly-diminishing forests. At night, cattle, sheep, and pigs would have been kept in pens in clearings using wooden rails cut from felled and living trees to try to keep wolves and bears at bay. By day, the farmers might have cut branches so the livestock could feed on the leaves, and the animals were possibly encouraged to graze more widely, accompanied by shepherds to protect them from predators.

Some farmers grew beans and peas; others grew flax which they made into linen for clothes. Dogs were used to herd sheep and cattle and worked as watchdogs; they were sometimes used to hunt deer, although others might have been treated as family pets as they are today. Cattle and goats provided meat and milk and sheep gave wool for garments.

Keeping their options open, our earliest farmers also retained something of a hunter-gatherer lifestyle for many years; they hunted deer and boar for meat, and deer antlers were often used to make primitive ploughs which scratched the soil, allowing seeds to be sown. They also trapped birds, gathered nuts and berries, dug for roots in the forest, and caught fish from streams. Gradually though, the pastoralist way of life took over, as it provided a year-round supply of food and didn't require such a nomadic existence.

Over the next 2000 years or so, these settlers spread across Britain. They grew wheat and barley for flour in small plots near the family home, which would have been built on cleared land grazed by their livestock. In the north of Britain they probably cleared trees and burned the more heathy vegetation from time to time in order to encourage fresh heather growth for livestock to feed on.

By the start of the Bronze Age, around 2000 BC, large areas of the countryside were laid out in unenclosed fields, usually square in shape. The primitive wooden ploughs (which quickly wore out) scratched the soil into small ridges rather than ploughing it, meaning that the soil had to be worked in two opposing directions in order to give a decent area of turned over soil to plant crops. Boundaries between some fields began to be marked by structures such as stone walls, hazel fencing, or hedging. Tracks and ways across the countryside allowed trade and the exchange of animals which helped prevent inbreeding.

Wildwood clearance continued apace, especially on the better, deeper, and more productive soils. By this time forest probably covered maybe 40% of the country. The physical work required to achieve this amount of forest destruction 'by hand' and by a small population is almost impossible to imagine. Some trees could be burnt, pines in particular; others could be felled with stone axes, far

Did Britain's wildwood look like this?

more efficient cutting tools than they might seem. Tree stumps were perhaps cleared by lighting fires around them or by making sure that livestock repeatedly wore away any regenerating shoots. But it was a huge task.

Then came a revolution! Although the wheel had been invented between 4000 BC and 2000 BC in several parts of the world at different times, wooden carts using wheels that could be horse-drawn or pushed by hand weren't in use in Britain until perhaps 1000 BC. Farm mechanisation had begun; a process that would eventually lead to the internal combustion engine and farm tractors.

By 700 BC there were maybe three million people living in Britain, the majority engaged in or supporting farming. The range of crops grown had widened; wheat and barley were still important but oats, broad beans (a hardy legume that returns fertility to the soil), vetches (as animal fodder), peas, rye, and flax were common too. Crop storage was developed, either using pits or raised stores to prolong a crop's usefulness. Sheep, goats, cattle, and pigs were often joined by poultry, geese, and ducks. Horses were a new arrival in the farmsteads, although they were not used for work so much as status symbols!

By 500 BC half of the original wildwood had been cleared. Farming typically revolved around small hamlets and farmsteads, each having areas of pasture for livestock and arable land for crops; woodland supplied timber and poles for penning the animals. Come the Iron Age and the use of iron ploughs, farming became more efficient, and a two year rotation was introduced. Crops were grown one year followed by the land lying fallow for a year with livestock grazing so that it recovered some of its fertility. This lead to surprisingly high crop yields and fuelled population growth. Britain was also exporting corn and cattle hides to continental Europe in exchange for wine and olive oil.

By the time the Romans arrived in 43 AD and moved north through the country over the next century, the wildwood was substantially depleted, more so in southern Britain where conditions were particularly favourable for farm development. What woodland remained in the lowlands was being managed as coppices in which trees were cut back on a rotation to re-grow and produce straight stems that could be used on farms.

WHAT DID THE ROMANS DO FOR FARMING?

'What have the Romans ever done for us?' asks Reg in the film *Monty Python's Life of Brian*. In addition to 'roads' and 'medicine', the unexpectedly positive responses include 'irrigation' and 'wine'. In reality, the Romans introduced far more than that in Britain. Improved ploughs, scythes, and even a corn harvester were just the beginning of their improvements to farming. The Romans ate well and fed their army well; they were used to a much more varied and healthy diet and they made sure that Britain's farmers got the culinary message. During the Roman period a range of new crops such as cabbages, onions, parsnips, carrots, leeks, and asparagus were first planted, plus herbs including thyme and rosemary.

They introduced higher yielding grains so bread became a more important part of the British diet. Walnuts, Sweet Chestnuts, and a larger variety of fruit trees including cultivated apples (rather than the native bitter crab apples), mulberries, and cherries came in too. Grape vines were planted on south-facing slopes. New farm animals were introduced: White Cattle and Guinea Fowl, together with Brown Hares and Pheasants for hunting. Maybe the rabbit was introduced too, although controversy still rages over whether they were brought here much later by the Normans!

The Romans developed England and Wales into one of the main corn-exporting regions of the then-known Western world. The demand-led economy that developed helped push agricultural output to new peaks as farmers started selling their produce in the expanding Roman towns and to the Roman army. The market economy was underway.

But the market economy started by the Romans was not to last. Towards the end of their occupation Britain's economy was in decline, compounded by the final withdrawal of the Romans in 410 AD. War, disease, and political upheaval under the Saxon rule that followed, along with emigration, all caused the population to fall.

Farming in Saxon Britain was done in large open fields, usually three to a village, and divided into regularly ploughed strips of barley, wheat, rye, pulses, or pasture, some of the latter cut for hay. Cattle and sheep were often brought in to feed on crop stubbles after the harvest; they were moved from other land, often heaths and moors that were used as common grazing in summer. Though modified over the centuries, and much reduced in extent, Laxton in Nottinghamshire is the only remaining working 'three open field' farming village in Europe still operating this ancient system.

Rural people – most of Britain's population was still rural – continued to be at least partly reliant on produce they could collect or hunt in the bits of forest that remained: berries, fungi, birds' eggs, fruits, and wild game. Even small birds were trapped, cooked, and eaten; they were a welcome supplement to a spartan diet. Forest produce could even make the difference between life and death in years when harvests failed. With such a precarious existence there was zero tolerance of vermin such as rats and mice that would otherwise eat stored farm produce essential for winter survival. There was also zero tolerance of any wild predators – wolves, bears, foxes, and others – that would kill livestock or wild animals such as deer and elk.

With a cooler and wetter climate, grazing of sheep and cattle in the hills and uplands – common in Roman times – diminished; scrub and trees started to re-invade over large areas. In the lowlands, peasant farmers typically cultivated an area of land sufficient for just one or two cows and would have kept a few pigs, goats, and sheep around their small farmsteads. The country went back to something more akin to a subsistence economy, quite a change from the heady days of Roman Britain.

Anglo-Saxon rule was replaced by that of the Normans after 1066, and the Domesday Book that followed their invasion provides a unique record to assess society, agriculture, and the countryside at large. The population was increasing again, and in the countryside small farmsteads grew into villages; today's more familiar lowland landscape of villages, manor houses, and churches took shape. Cultivated land totalled about three and a half million hectares (compared with around six million hectares today) with grain being sold to local towns and exported. The Domesday Book shows that England consisted of just 15% woodland and wood pasture (scattered trees on grassland); the days of the wildwood were well and truly gone. The remaining woods stayed much the same from the 11th century to the 20th. Almost all of this woodland in the lowlands was coppiced. Hedges, often planted using hawthorn, were more common in western areas or where woodland cover was greater rather than in places where the open field system of farming was still practised.

An increasing number of sheep were kept for wool; there were maybe three million by 1100 (there are 23 million today, excluding lambs).[iv] Wool was exported and sold locally for cloth. The Normans managed rabbit warrens commercially for the next 800 years. In consequence, rabbit meat was cheap and widely available. Many of the larger warrens were located on sandy soils (where digging was easy), on heaths, and in coastal dunes because such places were largely unproductive agriculturally. But rabbits are rabbits; they soon escaped and became widespread in the lowlands where they were frequently killed and eaten by local people.

By 1300, oats were grown widely in the uplands and rye was popular in drier areas. Wheat remained the premium bread crop while barley was grown for malt and beer. Flax and hemp were beginning to be grown again. The sheep flock had soared to nearly 20 million animals and wool was the UK's most important industry throughout the Middle Ages. British wool was sought after across Europe, a market that didn't decline much until the availability of cotton in the 19th century. Sheep faeces played an important additional role in the maintenance of fertility on lowland farms. Without the fertilisers available today, only by manuring with farm livestock and growing nitrogen-producing crops such as peas or beans could crop yields be maintained.

In the hills and uplands, with a warming, drier climate, vegetation burning together with sheep grazing ensured that the landscape was treeless. That pattern of land management still exists over the huge area of our upland, heather-clad moors, grassy or bracken-covered mountain slopes, and the high peaks themselves. It is a landscape we recognise and admire today yet it is undoubtedly very unnatural, not manmade but much modified by farming.

In the early 14th century, horses became much more commonplace and replaced oxen; they were faster at ploughing and at transporting goods by cart over longer distances, allowing the exploitation of new land far from existing markets. With a growing population, more farmland was needed. Most forest that provided decent crop-growing or livestock-rearing land had already been cleared so farmers and landowners turned to draining marshes and burning heathland and upland moors.

But not for long! The 14th century was arguably the most difficult time in this country's history, and it would be a very long time before Britain was to recover.

The survivors of the Black Death were to witness considerable change in the countryside. The smaller population's reduced demand for grain led to marginal arable land being converted to pasture (which required far less labour) or reverting back to scrub, woodland, and moor, some of which remains today. Although the effects of the disease were catastrophic, drastically reducing the country's workforce, its aftermath provided new opportunities for the survivors, particularly the poor. Peasants were able to get better paid jobs because labour was in such short supply; they married later, and had fewer children. They were also liberated from their family and geographic ties as the availability of work provided social mobility. Predictably, though, society's elite responded with controls and taxes, upping the tension in society.

DEATH AND DESTRUCTION

The great famine of 1315–22 and the Black Death (bubonic plague) of 1348–50 decimated Britain's population and radically altered the face of British agriculture. During the great famine – possibly the result of the volcanic eruption of Mount Tarawera in New Zealand and the worldwide climatic disruption it caused – arable crop yields fell dramatically. Prolonged wet summers prevented harvesting and spoiled crop quality. Fodder crops were affected too, with much hay being lost or even left uncut. Prices soared and much of the rural population suffered malnutrition, starvation, disease, and death. At times the situation was so desperate that children were abandoned, cannibalism increased, and future crops were forsaken as the seed corn was eaten during winter.

Spread by flea-infected rats, as well as by individuals who had been infected on the Continent, bubonic plague swept across Britain and reduced the population from about six million to roughly half that in under two years. Entire communities were wiped out, many towns and villages were abandoned, and population levels didn't again reach those of 1300 until some three centuries later. Fields were uncultivated, crops were unharvested, and livestock unattended.

Between 1361 and 1362 the plague returned, this time causing the death of another fifth of the population. It continued to return intermittently throughout the 14th and 15th centuries, sometimes only locally. Its impact became less severe, largely due to conscious government efforts to limit it; one of the last outbreaks of the plague in England was the Great Plague of London in 1665–66.

The Winchester Pipe Rolls – the most complete set of manorial accounts (from 1208 to 1710) in Britain – provide, in the most minute detail, a record of income and expenditure across the Bishop of Winchester's estates in southern England, the richest episcopal estate in the country. At the time of the Black Death they have numerous entries recording the rent received from tenants as: 'nothing because he is dead'!

In 1381 the effects of socio-economic change fired the Peasants' Revolt, a frustrated response to a poll tax. Feudal society was under pressure and peasants began to negotiate contracts with increasingly powerless lords of the manor. These new farmers, a force in the countryside, paid rents and specialised in different crops or livestock; as the consolidation of farms began, so the practice of enclosing land followed. It was the start of a process that would continue through several centuries, resulting in the creation of tens of thousands of miles of hedgerow, a man-made habitat of enormous wildlife value that has become an abiding feature of the lowland British landscape.

A century later there was a new development which, at least temporarily, preserved the remaining woodland. Well before the so-called Industrial Revolution, iron-making blast furnaces began to operate, the first in 1491 in the Weald of Kent. More were constructed, especially in Kent where there was a good supply of managed woodland to produce the huge amounts of charcoal needed to fire them. By 1600 they had spread country-wide and caused large areas of woodland to be coppiced. They functioned until the 18th century when it became cheaper to import iron.

It was after, and partly because of, the dissolution of the monasteries by Henry VIII's Government in the 16th century that managing forests and farmland became increasingly mutually exclusive aspects of land management. In monasteries across the country, the monks had grown rich on their large estates where the close association between forest and farm management had continued. Come the dissolution (1536–1541), most of the monastic land was split up and sold to provide landowners – who often cashed in the value of their forest as timber, while tenants were more confined to farming. Unlike the mixed forest/farm tradition still common today across much of the European mainland, in the UK farmers rarely manage their own woodland and have lost the skills associated with forest management.

Climate, as always, was a major factor in determining the quality and quantity of annual crop harvests, and throughout much of the 16th, 17th, and 18th centuries, winters were remarkably long and cold. Growing seasons were therefore shortened and crop yields reduced during the brief summers.

It was not until the 17th century when sophisticated Dutch drainage techniques were introduced to the extensive East Anglian fens that any large scale wetland drainage began in Britain. Although the drainage was slow to progress and to be adopted in other parts of the country, within a few hundred years the great wetlands of Britain – such as the Cambridgeshire fens and Romney Marsh in Kent – were no more. They have long since become fertile farmland.

Agriculture boomed in much of the 17th century as grain prices increased. Improvements in transport, particularly along rivers and coasts, brought beef and dairy products from the north of England to London. Arable yields rose considerably in the 1700s and Britain became a net exporter of cereals, the result of increased areas of ploughed land coupled with the maintenance of higher levels of stock and soil fertility. John Small applied mathematical calculations and science to the plough shape and the first cast-iron plough made its debut in 1763. The four-course crop rotation assisted livestock farmers too. Whereas winter rations of hay and straw were adequate to keep only a few stock year-round, higher yielding root crops allowed larger numbers to be overwintered. Livestock could be slaughtered throughout the winter and fresh meat was available most of the year.

COMETH THE HOUR, COMETH THE MAN

Often said to be the father of the Agricultural Revolution in Britain, Jethro Tull (1674–1741) was an English lawyer and small landowner from Berkshire, influenced by the early Age of Enlightenment. One of the first proponents of a scientific approach to agriculture, he helped transform agricultural practices by inventing or improving numerous implements. He is remembered in particular for inventing a horse-drawn seed drill that sowed seeds economically in neat rows. Prior to this, crop seed had been scattered randomly into furrows by hand. This was incredibly inefficient; in some areas high distribution rates wasted seed while in others there was inadequate seed for a crop. The seed drill dispersed seed evenly into ploughed furrows and closed the furrow up, a process that created a more regular and higher yielding crop. He later developed a horse-drawn hoe for eliminating weeds, a process that until then had been done manually with a hoe rather like a gardener uses to weed between a few lettuces.

A contemporary of Tull's, 'Turnip' Townshend is rather less well known. Charles Townshend (1674–1738), the Second Viscount Townshend, was a Whig statesman with a strong interest in farming turnips. In the early 1700s he popularised crop rotation as an alternative to growing the same crop year after year on the same land, a process that exhausts soil fertility and depletes yields. He suggested dividing land into four fields within which wheat, clover, barley, and turnips were grown in succession.

The clover and turnips 'renewed' the soil when grown after wheat or barley, clover being a nitrogen-fixing plant (unknown at that time) which incorporates essential fertility into the soil. Turnips and clover were also fodder crops; when the animals were let in to graze them, their droppings added more fertility. Townshend's method became known as the Norfolk Crop Rotation and was adopted widely. Sadly, the only thing Townshend earned for this wonderful discovery was his rather unfortunate nickname!

Two other men also figured importantly in the Agricultural Revolution. Robert Bakewell (1725–1795) a farmer and well-travelled agriculturalist from Leicestershire, together with the often-overlooked Joseph Allom from the same county, are today recognised because they were the first to implement selective breeding of sheep, cattle, and horses. Charles Darwin described it as artificial selection and it was partly the inspiration for his theory of natural selection; he cited Bakewell's work as demonstrating variation under domestication in which methodical breeding led to considerable modification of the forms and qualities of his livestock.[v] Until then, medieval stockbreeding was a haphazard business, a bit akin to the union of nobody's son with anybody's daughter.

The commercial value of the wool trade for Britain in the Middle Ages is reflected in the woolsack being the seat of the Lord Speaker in the House of Lords. In the 14th century, King Edward III (1327–1377) commanded that his Lord Chancellor whilst in council should sit on a wool bale in order to symbolise the central nature and huge importance of the wool trade to the economy of England. It was largely to protect the vital English wool trade routes with continental Europe that the Battle of Crécy was fought with the French in 1346. Embarrassingly, an inspection in 1938 revealed that the woolsack was, in fact, stuffed with horsehair! It was hurriedly re-stuffed with wool from all over the Commonwealth as a symbol of unity.

But the importance of the wool trade was to diminish. Philip Walling tells the story of Henry Best, a yeoman farmer in Yorkshire, who, in 1641, sold his entire wool clip for that year for just over £11.[vi] At the time, employing a farmhand for a year cost £3. At today's much reduced price, the same amount of wool would fetch maybe £240 and the cost of employing a farmhand would be a hundred times that figure! Wool lost out to imported cotton in the 18th and 19th centuries, and to synthetic fibres in the 20th. Its price has never recovered and sheep are very largely kept today for lamb meat. However, recent years have seen an increase in value, but it is still a minor product of sheep farming. The average price paid by the British Wool Marketing Board to producers for a kilo of sheep's wool in 2014/15 was just over £1 (though top quality wool can attract much more) – about three times its value at its low point in 2006 – largely due to increased demand, in part from China.[vii]

The century between 1750 and 1850 witnessed a massive change in farming practices as the implementation of the Inclosure Acts brought about the demise of the open field system that dominated the countryside. Inclosure Acts for small areas of open land had been passed sporadically since the 12th century, and much enclosing was done by local agreement, but with the rise of the Industrial Revolution, they became more commonplace. Between 1604 and 1914, over 5,200 individual Inclosure Acts enclosed nearly three million hectares of land.

Many so-called 'wastes' – heaths in the lowlands, marshes, and moors in the uplands, all of huge value for wildlife – were enclosed and brought into cultivation. Wherever possible, marshes were drained by digging ditches through them. This 'common' land had been under a kind of collective control, usually used for common grazing, hay cropping, or crop rotation, often by landless peasants who were then driven to find work elsewhere. More and more land was enclosed with planted hedges or stone walls to create fields of varying shapes and sizes, overriding the ancient Saxon strips, thereby producing the pattern of countryside we are more used to today.

ON DISCONTENT

John Clare (1793–1864), the English poet, farmer, and countryman was appalled at the impact of the enclosures on village people. In his 1821 poem 'The Village Minstrel' he wrote:

There once were lanes in nature's freedom dropt,
There once were paths that every valley wound –
Inclosure came and every path was stopt:
Each tyrant fix'd his sign where paths were found
To hint a trespass now who'd crossed the ground:
Justice is made to speak as they command,
The high road now must be each stinted bound:
'Inclosure, thou'rt curse upon the land,
And tasteless was the wretch who thy existence plann'd

With much less open land, shepherding became increasingly obsolete since livestock could be penned in a particular field. There was no need to wander with your flock to find grazing. While Little Bo Peep no longer worried about losing her sheep and Little Boy Blue could give his horn blowing a rest, employment fell. Far worse, enclosure resulted in substantial rural depopulation because villagers were denied common grazing on the 'wastes' that were now enclosed. No longer could they dig fuel such as peat from wetlands, collect fallen wood, or cut heather for livestock bedding. Whole villages disappeared as their occupants left or died out; there was widespread poverty and misery. Many people regarded enclosure as a curse – it was certainly the most dramatic change to our countryside for a thousand years.

Board of Agriculture (a chartered society created in 1793) statistics suggest that there were over three million hectares of so-called 'wastes' in England and Wales in 1795, a fifth of the land area. Little of that remains today; it has long since been enclosed, drained, burnt, or cultivated. At the time such wastes were regarded as sterile, barren places. Daniel Defoe, the writer, journalist, and spy, wrote about one, Bagshot Heath in Surrey, when he visited it for his *Tour of the Whole Island of Great Britain*, published in 1724:

Here is a vast tract of land some of it within seventeen or eighteen miles of the capital city; which is not only poor, but even quite steril, given up to barrenness, horrid and frightful to look on, not only good for little, but good for nothing....the product of it feeds no creatures, but some very small sheep, who feed chiefly on the said heather, and are very few of these, nor are there any villages worth mentioning and but a few houses or people for many miles far and wide.

Today, Bagshot Heath is highly valued for outdoor recreation and for the wealth of plants and animals it sustains, many of them uncommon.

Before 1800, farmers started to develop regional specialisations – most of which remain largely unchanged today – and mixed farming began to decline. Eastern and southern England, for instance, with flatter fields, productive soils, and a warmer, drier climate concentrated on cereal growing. The uplands of the north and west of Britain are better suited to extensive sheep grazing, South West England for dairy farming, and so on.

Traders in the late 1600s were provided with bounties for exporting rye, malt, and wheat (all classified as corn at the time), and the same commodities were taxed when imported into England. The Corn Laws introduced between 1815 and 1846 imposed further restrictions and tariffs on imported grain. Designed to keep grain prices high and favour domestic producers, they boosted home production and encouraged the farming industry to invest and innovate and bring yet more uncultivated land – heath, marsh, and grasslands – into cereal production.

But this early farm protectionism was opposed by many politicians and by the growing urban population who had to buy its products at higher prices. Repealed between 1846 and 1849 through a gradual reduction of the tariff, Britain's wheat industry was largely destroyed as cheaper American imports, of cereals especially, took over. By 1891, reliable refrigeration brought cheap frozen meat to Britain from Australia, New Zealand, and South America.

FARM POWER

The first powered farm implements started appearing in the early 19th century: portable steam engines on wheels that could be used to drive mechanical farm machinery by way of a flexible belt. They were towed from farm to farm by horses. A farming revolution, they replaced arduous, labour-intensive tasks such as corn threshing.

It was a relatively short step to the first traction engines – steam engines on wheels that could be driven around the farm to undertake a variety of tasks. Richard Hornsby & Sons are credited with producing the first oil-powered engine tractor in Britain in 1896. It was invented by Herbert Akroyd Stuart. Gaining popularity in the 1910s, when they became smaller and more affordable, Henry Ford's Fordson became a wildly popular mass-produced tractor soon after. By the 1920s, tractors with gasoline-powered internal combustion engines had become the norm and were already beginning to nudge out the need for so much employed labour and the horses no longer needed to pull machinery.

An agricultural revolutionary: 1936 Fordson Model N tractor
(courtesy of Markus Hagenlocher).

The great agricultural depression was the result. Lasting for nearly 30 years, there was significant rural depopulation and, in an increasingly urban society, few spoke up for agriculture. While the countryside started to become a remote and distant environment for an industrialised culture (as it remains today), the railway boom of the late 19th century provided the means to transport farm produce to towns and cities quickly and efficiently. The milk industry burgeoned; collected by lorry in churns from farms, it could be transferred quickly to cities by rail tanker. The rail transport network started to be superseded by road transport in the 1920s and 1930s giving more flexibility and enabling the gradual introduction of chilled tankers.

The drastic slimming down of farm employment was underway. It has never stopped since. Office for National Statistics figures show that today, less than 1% of the UK workforce is employed in farming; in 1841, it was 22%.[viii]

Spending on land drainage, buildings, machinery, and roads linking to the railways all fuelled the Agricultural Revolution. Technological developments took place across the whole spectrum of mechanisation, with reaping and threshing machines, new ploughs, and drills increasing in popularity. Huge quantities of nitrogen fertiliser in the form of seabird droppings (guano) were imported from South America by ship and widely used by the mid-19th century.

World War II caused enormous changes on farms. For our island nation, having to import a substantial proportion of our food meant much of our provisions were subject to U-boat attacks on freight ships, creating food shortages. The 'Dig for Victory' campaign pressed farmers into ploughing up pastures and meadows and draining as much marginal land as possible to grow crops. The UK had to become as self-sufficient in food as it could. Farming was no longer in the doldrums; it was essential to victory.

When it was over, the 1947 Agriculture Act reformed British agriculture. Never again would Britain become this close to starvation because the country could be blockaded. Before World War II, Britain was importing 55 million tons of food a year; by the end of 1939, this had dropped to 12 million tons and food rationing was introduced at the start of 1940. It did not completely end until July 1954.

The Act guaranteed minimum product prices, markets for farm goods, and tenure for farm tenancies; a farmer could be assured that his land would not be taken away and whatever he grew would be sold at a known price. This and the 1948 Agricultural Holdings Act made it harder to evict tenant farmers; over the long term it resulted in more owner-occupied farms. The post-war emphasis was on greater production using fertilisers, better combating of pests and weeds with pesticides, and higher-yielding crops, grasses, and livestock through improved plant and animal genetics. It was a success. In 1938–1939 farm output in Britain was valued at £2.5 million; in 1951 it was over £100 million. In addition, the legislation provided farmers with a degree of prosperity and security not known since the mid-19th century.

Chemically synthesised fertilisers containing the main plant nutrients (nitrogen, phosphorus, and potassium) started becoming available in the first few decades of the 20th century. Today, together with pest control, they are vital for sustaining crop yields and have the advantage of allowing the same crop to be grown on the same land year after year. The use of nitrogen-based fertilisers increased from around half a million tons per season in the 1960s to over 1.5 million tons in the late 1980s but has decreased steadily since to less than a million tons today. According to the Agricultural Industries Confederation, there's more focus today on developing crops that require less fertiliser (due to its cost and runoff causing water pollution) and more use of manures.

Many people assume, wrongly, that pesticides are a very modern invention. But their earliest recorded use was of powdered sulphur (still permitted, rarely, on UK organic farms) about 4,500 years ago in Ancient Mesopotamia. Losses of crops to pests were so significant that steps were taken even then to try to reduce it. All early pesticides were natural substances, many derived from plants, animals, or smoke from burning fires used against mildew and fungal blights. Weeds were controlled mainly by hand-weeding and hoeing, although various 'chemical' methods are also described, such as the use of salt or pyrethrum which is derived from the dried flowers of a *chrysanthemum*.

In the 19th and 20th centuries, by-products of industrial processes provided organic compounds which were used to kill fungal and insect pests, whilst synthesised ammonium sulphate and sodium arsenate were used as herbicides. The drawback for many of these products was that they needed high rates of application, they were not selective in what they killed, and they were generally toxic.

It was in the 1940s that synthetic pesticides really got going with the effective and inexpensive DDT, aldrin, dieldrin, and others. Throughout most of the 1950s, consumers and most policy makers were not overly concerned about the

Today's crop yields are at an all-time high: oil seed rape in Shropshire.

potential health risks of using them; food was cheaper because of the new chemical formulations and there were no documented cases of people dying or being seriously hurt by their normal use. The 1970s and 1980s saw the introduction of the world's greatest selling herbicide, glyphosate, as well as a wide range of other herbicides and insecticides. However, pest resistance to some of these chemicals had started to become an issue.

Research in the 1990s concentrated on finding variants of existing pesticides with greater selectivity for particular pests but with lower human and environmental toxicity. Many of these new agrochemicals are used at a scale of grams rather than kilograms per hectare.

Fertilisers, pesticides, and increased mechanisation providing hugely increased yields of crops, meat, and milk are major components of Britain's recent agricultural success. If today's arable farmer was to grow a crop of wheat in the unlikely setting of the Wembley Stadium, he could produce enough to make the bread for one sandwich for everyone in a capacity crowd of 90,000 people! Wheat yields have risen enormously, from an average of less than two tons/hectare in 1892 compared with eight tons/hectare today.

By no means all farmers have gone down the intensification route of high yields facilitated by fertilisers and pesticides. Organic farming shuns chemically synthesised fertilisers (but not natural ones such as manures), most pesticides (except a few naturally-occurring chemicals), genetic modification, and the routine use of drugs and antibiotics. Instead, it relies on crop rotations and other forms of husbandry to maintain soil fertility and control weeds, pests, and diseases. Organic standards don't allow intensively-housed livestock or systems where a large amount of total feed has to be brought in off-farm; pigs and poultry must be managed with extensive outdoor access. In essence, modern organic farming is akin to farming as it was done around the turn of the 20th century when chemical fertilisers and modern pesticides weren't available.

Financial incentives were introduced in 1994 to encourage farmers to convert from non-organic (called 'conventional' farming by those who are non-organic) to organic. The aim was to produce higher-value foods for more discerning consumers willing to pay a premium price and get, as they believed, healthier food. The number of UK organic farmers reached a peak of 5,000 in 2008, although their numbers have since declined and there are about 3,500 today. The amount of land devoted to organic farming is also falling, and is much less than in several other EU countries (see Chapter 10).

JOINING UP

Another momentous change took place in 1973 when the UK joined the EU. The EU's Common Agricultural Policy (CAP), fraught with controversy because it consumes well over a third of the entire EU budget, set a minimum price for almost all farm produce (see Chapter 4). They also defined so-called Less Favoured Areas (LFAs) of member states in which direct payments were made to farmers based on the numbers of livestock a farmer kept. These 'headage payments' were effectively social subsidies 'disguised' as support payments for farming. They have proved essential in keeping people farming sheep in our hills and uplands. For many years farmers were paid per head of livestock, which encouraged overstocking of the land.

During the late 1970s and 1980s in particular, substantial capital grants funded by the EU encouraged a massive flurry of intensification of farming in the LFAs in particular. New farm access roads were driven through hills and moors, often scarring the landscape; expensive drainage schemes were installed in wet fields and marshy ground; scrub was cleared to provide yet more pasture; and existing permanent pasture and hay meadows replete for centuries with an array of wild flowers were ploughed up and sown with a cultivated rye grass-clover mix. This was all to graze yet more sheep and claim yet more headage payments. It was bonanza time for farmers and farm contractors alike.

CAP's costs soared with heavily subsidised food production, and resulted in huge over-production and storage requirements – the infamous 'butter mountains' and, outside the UK, 'wine lakes' the contents of which had to be sold off outside the EU at a loss. Overproduction was causing considerable environmental damage too. In 1987, farmers began to be paid for a limited amount of habitat and wildlife protection through 'agri-environment' schemes, derivatives of which are central to the CAP today. Gradually, the links between farm production and subsidies were removed.

Other changes have taken place in recent decades too, some much more subtle than others. A major one has been a substantial replacement of hay making with silage making to provide feed for cattle and sheep during the winter months. Hay making in Britain, especially in the wetter west and north of the country, was always very hit and miss as the hay crop had to be cut in midsummer when the plants were reasonably dry, thereby improving its chances of drying thoroughly for storage. Damp hay decomposes and this bacterial process generates enormous heat which can ignite the stored crop. Silage making is much easier; it involves cutting and collecting grass from a pasture, usually a few times in a growing season – it doesn't need to be as dry for silage making – and storing it under anaerobic (air free) conditions where it ferments before being usable as a feed. Many traditional hay meadows have now been ploughed up and sown with a productive ryegrass-clover sward, obliterating its flowering plants.

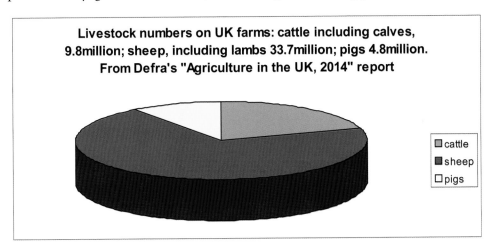

Livestock numbers on UK farms: cattle including calves, 9.8million; sheep, including lambs 33.7million; pigs 4.8million. From Defra's "Agriculture in the UK, 2014" report

- cattle
- sheep
- pigs

In the lowlands, another substantial change over the last few decades has been the trend away from sowing cereal crops in spring to sowing them in autumn instead. Using varieties of cereals bred to grow, albeit very slowly, through the cold damp winter, they produce higher yields and an earlier crop. It might seem a minor change in arable farm practice but sowing the crop in autumn requires ploughing the stubble of the previous crop into the soil before sowing the seed rather than leaving it all winter on the soil surface, where flocks of finches and sparrows find spilt crop seed and weeds. In the last year or so, leaving stubble over winter has become more commonplace in order to abide by new EU rules to prevent soil runoff into watercourses.

Intensive indoor rearing of poultry and pigs started slowly but picked up pace in the 1960s. Driven by economies of scale and consumers wanting cheap meat and eggs, the size of these operations has grown enormously. Public concern about animal welfare has led to adjustments in how such industrial farming units are run, and further changes seem likely. The quantities of manure produced can be enormous, and recent EU regulation has been focussed on ensuring that such effluents don't pollute nearby watercourses and that emissions of polluting and climate-warming gases are minimised.

Highly efficient though modern farming in Britain most certainly is, its contribution to the economy as a percentage of GDP has fallen to less than 1%. We are less than 60% self-sufficient in foodstuffs; the rest we import, although some of that is to satisfy consumer demand for 'out of season' fruit and vegetables.

Over the last decade or two, many farmers, maybe up to half, have diversified into other part-time occupations. Encouraged by successive governments, many now run small tourism ventures (especially farm B&B) with lesser numbers going into renewable energy generation; more novel crops such as herb production; fish farming and recreational angling; ice-cream making; bottled water production; and much more besides. Start-up funding is often available, along with a considerable amount of advice, both economic and technical. Many, too, have added value to their own produce with on-farm cheese, yoghurt, and cream making or by selling their own meat and other products direct to consumers.

Our food preferences have altered too. Unlike our meat consumption prior to and for a couple of decades after World War II, most of today's British meat consumers shun offal – brawn, chitterlings, faggots, liver, and heart – even if some leading chefs now promote it. So current livestock is bred to have smaller rumens (the main vegetation digestion organ in sheep and cows) making them less efficient digesters of rougher grasses. Farmers have responded by replacing traditional meadows with fast-growing, easily digested grass leys as well as silage rather than hay for winter feed.

Britain's land today would be unrecognisable to our early farmers; just 13% is covered in woodland and, of that, only 4% of it is the native broadleaved woodland our early settlers would have any hope of recognising. Most other EU countries retain far more woodland: Germany and France, for instance, are over 30% wooded, a similar proportion to the US. While much of Britain's upland and hill farming would be recognisable to a Saxon farmer, our lowland agriculture – especially the intensive growing of crops such as wheat, barley, rapeseed, and maize – has become industrialised in ways they couldn't have even imagined; farming in our lowlands has metamorphosed into agri-business. It is far removed from the practices of farming in even the mid-20th century, and totally alien to Britain's early farmers who had to cope with backbreaking labour, dangerous predators, poor soil tilling, and catastrophic crop pests and diseases. They certainly didn't have a Common Agricultural Policy or any other supports.

Endnotes

i   'The Origins of Agriculture in the Near East,' Melinda A. Zeder, 2011. Current Anthropology, vol 52, no S4. University of Chicago Press.

ii Trees and Woodland in the British Landscape, by Oliver Rackham. J.M Dent and Sons Ltd, 1976.

iii 'How fragmented was the British Holocene wildwood? Perspectives on the "Vera" grazing debate using fossil beetles,' N. J. Whitehouse and D. N. Smith, 2010. Quaternary Science Reviews 29 (3-4): 539-553.

iv 'Livestock numbers in the UK,' Defra, 2016.

v   On the Origin of Species, by Charles Darwin. John Murray, 1859.

vi Counting Sheep, by Philip Walling. Profile Books, 2014.

vii British Wool Marketing Board statistics, 2015.

viii '170 Years of Industrial Change Across England and Wales,' Office for National Statistics, June 2013.

CHAPTER 3

DRAINING THE LIFEBLOOD

You cannot get through a single day without having an impact on the world around you. What you do makes a difference, and you have to decide what kind of difference you want to make.

JANE GOODALL, BRITISH PRIMATOLOGIST, ETHOLOGIST,
ANTHROPOLOGIST, AND UN MESSENGER OF PEACE

Since the first farmers arrived on our shores, Britain's countryside has been in a state of constant change. By the time the Normans invaded in the 11th century, the land they conquered had lost the lion's share of its original wildwood, the natural forest that had once clothed at least two-thirds of Britain. While we quite understandably concern ourselves today with the impact of modern agriculture on a relatively small number of birds or the decline of many of our bees, the loss of our wildwood was an infinitely greater wildlife disaster. Over the last 5,000 years or so, Britain has lost an area of woodland larger than the whole of Greece: over 150,000 square kilometres.

Britain's wildwood was the richest wildlife habitat ever to occupy our land. It was a veritable cornucopia of plants, from flowers and grasses to mosses, ferns, and liverworts. It was home to the widest variety of animals ever to exist here, the likes of which we have never experienced since.

Today, just 4% of our land is broadleaved woodland, roughly 10,000 square kilometres. Maybe half of that has been planted in more recent times; the other half or even less, historical ecologists tell us, is known to have existed since about 1600, when trustworthy maps first became available (1750 in Scotland). These areas might well be the last remnants of what was the wildwood, and have been dubbed 'ancient woodland'. Covering 4,000 square kilometres or so, albeit dotted around the country in small parcels and often confined to steep or other inaccessible places, they have been much modified over the millennia. People have been felling parts of it, re-planting some areas, introducing different trees, and much else. But it's our link with the wildwood.

The extent of the wildwood, its huge variety across the country, and the millennia over which it existed, mean that two-thirds of our breeding land birds, half or more of our butterflies and moths, and at least one in six of our flowering plants are entirely or partly dependent on woodland. Their numbers must have dwindled considerably as their habitat disappeared.

WILD WOODLANDS

Judging from the variety of ancient woodland fragments that still exist, we can safely assume that the composition of the wildwood varied enormously across Britain: oak and birch woodland on many acid soils in the west, alder woods on wetter ground, ash woods on more neutral and alkaline soils, lustrous Scots Pine woodland, and much more. Some of it was probably dense with understorey shrubs – holly and hazel perhaps –and climbers such as ivy and clematis. Other areas were probably more open, the result of occasional fires or heavy grazing by wild mammals, thereby allowing heaths or grassland to develop.

Dense forest would have encouraged shade-loving plants such as Sanicle and Wood Spurge to prosper; spring-flowering gems like Bluebells too. If the remnants of oak woodland remaining in the very west of Britain reflect the composition of the wildwood there, wetter conditions would have favoured a huge array of moisture-loving mosses, liverworts, and ferns, plus invertebrates that depend on such plants. Where understorey shrubs were dense, Nightingales would have filled the air with their fluted whistles and ratchet-like rattles while woodpeckers hammered out nestholes each spring. The quiet calls of chequered Pied Flycatchers and tiny olive and yellow Wood Warblers singing their metallic, shivering trills, both migrants from Africa coming here to breed in summer, would have been drowned out in the cacophony.

The likes of badgers, otters, foxes, wildcats, and hedgehogs made it here before we were rendered an island nation when the sea flooded across the land bridge with the Continent in about 6,000 BC. Red and Roe Deer already inhabited the wildwood; Aurochs roamed the woods too, as did Elks, Grey Wolves, European Lynx, and even Brown Bears, Wild Boars, and European Beavers.

Our first farmers, scraping a living by growing a few crops and grazing some sheep, goats, and cattle that they had domesticated, couldn't afford to tolerate predators killing any livestock. Bears and wolves were therefore hunted mercilessly. Other mammals like deer and boar were hunted for their meat and antlers; their bones were used as tools and their skins for bedding and clothing.

As the numbers of farmers increased, and more and more of the wildwood was cleared, many of these mammals didn't have a great future. No one is sure when Aurochs breathed their last within our shores, but they have probably been gone for the best part of two millennia, possibly longer; today they are extinct worldwide. Elks had already disappeared during this time. Eurasian Lynx were extinct here by the 6th or 7th century and Brown Bears disappeared sometime between then and the 10th century. Wolves hung on until the 15th century, later in Scotland but only in the remotest places. Beavers were gone from Britain by the end of the 16th century, trapped for their fur and castoreum, a glandular secretion they used to mark their territories which, unfortunately for them, people believed had medicinal properties.

Wild Boars didn't last as long as the 16th century. Hunted for their meat, and despised because of the damage they caused by rooting in farmed land and the possibility of their hybridising with domesticated pigs, their future was cut very short. Britain's only wild member of the cat family – rather prosaically called the Wild Cat – was thought to have been common and perfectly adapted to hunting in the wildwood. Hunted in turn, and deprived of their habitat, they disappeared from southern England in the 16th century. The last one recorded in northern England was shot in 1849 and today a small population (much inter-bred with domestic moggies gone feral) survives only in the Scottish Highlands.

It's ironic that the rapidly declining wildwood was a lifeline for early farmers, both in times of food shortages and in enhancing a very restricted diet generally. Berries, fungi, bird's eggs, and fruits were harvested and large numbers of birds and mammals were trapped. Anything from sparrows to wild duck would help feed a hungry family.

Naturally, as the area of wildwood continued to decline, the populations of woodland birds, mammals, plants, and invertebrates would have depleted with it. As more woodland was coppiced to provide much-needed poles for building and fencing, many woods developed an open, sunnier aspect, good for spring flowers and sun-loving insects, but bad for the species that needed shade to survive. For example, the numbers of butterflies and moths – such as Purple Hairstreaks and Blotched Emeralds – that depend on shaded woodland would have plunged. Those preferring sunny glades and open areas – like Dark Green Fritillaries and Brown Scallop moths – would have gained an advantage.

The birds of open lands that we are so familiar with today, such as Meadow Pipits or Skylarks, would have been unusual sights when the wildwood dominated. Presumably they were only to be found in very small numbers in the most open glades, created perhaps by summer fires or grazing. The clearance of wildwood and the domination of crop-growing land and pasture for livestock tilted the balance of woodland and open land birds. It was a one way flow. No longer would the drumming of woodpeckers and the mellifluous rattles of Nightingales fill the springtime air; instead, the sweet cadenzas of Skylarks high on the wing, the 'pitter-patter' calls of Grey Partridges taking flight from a grassy bank and the monotonous jangling songs of Corn Buntings became more common.

An estimate of some of the larger mammals in Britain when the wildwood dominated the land

Red Deer	1 million plus
Roe Deer	1 million plus
Elk	67,000
Auroch	99,000
Wild Boar	1 million
Grey Wolf	20,000
European Lynx	10,000
European Beaver	35,000
Weasel	423,000
Polecat	104,000
Stoat	50,000 plus

(Compiled from information published in 1999 by Dr Derek Yalden, a leading British zoologist and mammal expert, who based the estimates on current populations in Poland's extensive and ancient Bialowieża Forest.)

Where wildwood clearance gave way to heather, gorse, and bracken-covered heathland, birds such as Stonechats and pretty, dark grey and red wine-bedecked Dartford Warblers were to be found. No one knows if they were already present in very small numbers, or if they spotted an opportunity and spread across from continental Europe. Summer migrants, Corncrakes had probably become a bothersome breeding bird in arable crops, their rhythmic, grating call enough to keep people awake at night. Before the wildwood was cleared, these skulking ground birds were possibly not even in Britain; farmers had started creating an artificial habitat that suited them very nicely indeed.

On upland moors where, depending on their elevation and degree of waterlogging, tree cover might always have been more restricted, breeding waders such as Golden Plover, Curlew with their haunting calls, and tubby Red

Grouse had already settled in. They gradually gained more habitat as woodland on drier hill land was felled or burnt down and the land put over to sheep grazing. Some species probably adapted to the massive changeover of habitats better than others. Birds of prey such as Buzzards and Kestrels would have started hunting over more open ground, their phenomenal eyesight making it easier to spot prey where fewer trees hindered their view.

With the amount of open land increasing, it perhaps isn't surprising that the Romans most likely introduced Brown Hares to the UK for hunting and eating. For centuries they were a characteristic feature of lowland farmland in Britain until they declined substantially between the 1960s and 1990s, the main period of farm intensification and crop specialisation. Another introduced species, the Pheasant, was common here by the 15th century. According to the Game and Wildlife Conservation Trust, 35 million of them are now bred and released every year to be shot or, in many cases, die on roads or from winter starvation. Recent research suggests a link between the continuing annual increase in released Pheasants – together with the changeover in recent decades from spring sowing of cereals to autumn sowing (thereby ploughing in any spilt seed) – and the enormous decline of many small, seed-eating birds such as Tree Sparrows.[i]

Whether the regularly tilled, hoed, hand-weeded, and wide open farmland of Roman and Saxon times that replaced the wildwood was good for wildlife is hard to tell. When there were shortages of labour during periods of human population decline, it was presumably managed less intensively. That would have provided an opportunity for wild flowers and grasses to develop, encouraging insects to establish and birds to feed and breed in longer vegetation. When the human rural population was buoyant, there would have been little room for anything other than planted crops and zero tolerance of anything considered to be 'vermin'. So it was that many crows (including Ravens), birds of prey, moles, polecats, stoats, and others would have met an early death because they would have been considered to be in competition, directly or very indirectly, with those farming the land.[ii]

Cereal crops, though, would have been speckled with the then common 'weeds' such as purple-pink Corncockles, yellow Corn Marigolds, and scarlet poppies, and there was always some land lying fallow for a year where natural vegetation could re-establish. Small birds such as finches could pick up any spilt grain or 'weed' seeds after harvest as they combed the wheat and other cereal stubbles. However, assiduous 'vermin' trapping – encouraged by bounty payments introduced in the 16th century – and the killing of any birds or mammals worth eating meant that the open fields that came to dominate the lowlands were unlikely to have been anything approaching wildlife havens.[ii] In theory, they would have been ideal habitat for mammals such as Brown Hares, but a hare would be unlikely to last long before it became a much-needed family dinner. Breeding birds might have included Skylarks, Corncrakes, Quails, Grey Partridges, and maybe Stone Curlews; all of these, and their eggs, would have been at risk of becoming dinner too. Larks, commonly consumed with bones

*Not much of a wildlife haven. A Saxon open field system,
this one preserved at Laxton, Nottinghamshire.*

intact, have historically been considered wholesome and delicate eating; lark's tongues were particularly valued!

In the 1970s, when Dr Ernie Pollard and his colleagues studied the birds of the Laxton open field system in Nottinghamshire (see Chapter 2) they found that the range of open land birds characteristic of surrounding farms – where fields were instead separated by hedges – was absent.[iii] Only a few species such as Skylark, Lapwing, and Grey Partridge were to be found on Laxton's wide-open acres.

Roger Lovegrove, in his study of human persecution of British wildlife suggests that the introduction of rabbits to Britain in the 12th century (though maybe earlier) increased the pressure on natural predators in the countryside even more.[ii] Buzzards, Red Kites, Wild Cats, Red Foxes, and Polecats soon adapted to killing them and found themselves in direct competition with rural dwellers. Birds of prey would have had a better time hunting over the so-called 'wastes', areas of land used in common by the livestock of a village and located outside the cultivated fields. Consisting of rough grassland, maybe with a few trees, heathland, and marsh, in all probability these habitats did at least support decent populations of plants, insects, small mammals that could outwit the trapping and hunting, and birds. In times of human population decline – after the Black Death in the 14th century especially – trees would have re-established in such places, maybe even on some cultivated land, thereby re-creating a temporary wildlife habitat, because there simply were not enough people surviving in the countryside to do the cultivating or to shepherd livestock.

The upland moors that had been cleared of trees would also have reverted back to scattered tree and scrub cover at such times, again altering the balance of wildlife species occupying these habitats. But once the human population had recovered in the 17th century, the moors would have been burnt to kill off any trees and regenerate new growth from the rank vegetation for sheep to graze. This is the condition that much of our upland moorland is in today.

When the large open fields so characteristic of Saxon times started to be divided up with planted hedges, especially as a result of the Inclosure Acts in the 17th to 19th centuries, wildlife adaptation became necessary once again.

HEDGING THEIR BETS

Mature, thick, livestock-proof hedges with a narrow skirt of grassy, flower-filled turf along either side provide excellent wildlife habitat, especially the more ancient hedges which are richer in a variety of shrubs, flowers, and grasses. According to the conservation charity Buglife (motto: 'saving the small things that run the planet'), over 1,500 insect species have been recorded at one time or another living or feeding in hedgerows. Grass and flower seed supports Linnets; there are butterfly and moth caterpillars for Cuckoos; berries for Fieldfares and other thrushes; mice and voles for hunting owls; and much else besides.

Enclosing fields with hedges was a huge boost to farmland wildlife and created some alternative habitat for those plants and animals that could readily adapt to an open, sunnier 'woodland' substitute. Birds such as Yellowhammers, previously confined to the sunny edges of woodland clearings, adjusted well to the new woody lines. In these pre-pesticide days, birds breeding in the hedges or in the rank vegetation at their base, such as the Grey Partridge, could find plenty of flower seeds and small insects out among the crops or pasture in the adjacent fields.

With increasing areas of arable land growing wheat and other cereals from the 18th century onwards, Corn Buntings probably prospered too. Crops rich in flowering plants such as Corncockles, chickweeds, and thistles, plus a welter of invertebrates, surrounded by hedges for nesting would have been irresistible. Blackcaps – small warblers wintering in Africa and breeding in Europe – were perhaps always common breeders in the wildwood but soon adapted to hedgerows and their more scattered trees. Yet other migratory warblers never have. It's no good listening out for the metallic shivering trill of a tiny Wood Warbler in a hedge. They remain resolutely confined to the oak and birch woodland of the west of Britain. Whilst this woodland once clothed vast areas of land, today it is reduced to remnants on steep slopes where felling, burning, and heavy sheep grazing haven't been able to make major inroads.

Upon maturing, these hedges became narrow strips of something approaching woodland, tens of thousands of kilometres of woody, meandering habitat providing feeding and breeding places a-plenty. They might have been planted with just one species of shrub, hawthorn commonly, but over many years other shrubs would have seeded in, often carried by birds. Some of the original hedges were more diverse to begin with; they would have been planted using a mix of saplings from the wildwood or adapted from existing lines of wildwood trees and shrubs.[iii] No one knows how many kilometres of hedgerow were created this way, but Oliver Rackham – who described them as 'the ghost of woods that have since been grubbed out to leave only the edge of their field' – estimated that well over 300,000 kilometres of hedge were planted between 1750 and 1850 alone.[iv]

So it was that hedges, maturing over centuries and carefully managed by farmers to retain their value as livestock-proof field enclosures and effective windbreaks, became a valued part of Britain's lowland landscape. Farmers needed hedges, and so did a fair chunk of our wildlife. But it was not to last, at least not in its entirety. Since World War II, there has been a drastic loss of hedgerows, many of them grubbed out to create larger fields more appropriate for the burgeoning scale of crop production and the ever larger equipment used to sow, treat, and harvest. Others were lost through neglect as farmers failed to maintain them. In 1946 there were an estimated 800,000 kilometres of hedgerow in England; by 1993, over half had gone.[v] With them went many individual, often mature trees that had grown up in the hedges protected from grazing livestock. Our farmed countryside had suffered a double whammy.

Because of enormous concern from the public and from conservation organisations about such declines, Hedgerow Regulations became law rather belatedly in 1997. Designed to protect exceptionally species-rich hedgerows and those of landscape, archaeological, and historical importance (usually the oldest), consent for removing a hedge is now required from a local authority and is given only if it doesn't qualify as important. Criticism that the Regulations are fiendishly complex and seemingly imperfect in definition has not so far resulted in their amendment.

Although today far fewer hedges are grubbed out, many are still declining in wildlife value through neglect. Managing a hedge – so-called 'laying' – is a labour-intensive and skilled job that involves weaving partly-cut branches from the hedge horizontally through the upright stems of the shrubs. It keeps the hedge thick and healthy, without which its shrubs get overgrown, gaps appear at the bottom which livestock can exploit, and it ends up as a rather stark row of trees or a pitiful scattering of grazed shrubs before it dies out altogether. Too frequent trimming, often done by machine, can be equally destructive.

The considerable loss of hedges before protection was imposed in 1997 has undoubtedly caused a decline in a huge number of plants, invertebrates, small mammals (including Hazel Dormice), and farmland birds.

They came…and went. Thousands of kilometres of hedgerows have been grubbed out or fallen into disuse.

According to The English Hedgerow Trust, a charity promoting hedge conservation, 21 out of 28 lowland mammals and 23 out of our 54 butterfly species breed in hedges.[vi] In countryside with little or no woodland left, they are essential for the survival of much wildlife; more than that, in an era of climate warming when the distribution of species is changing as they adapt to warmer temperatures, flightless animals can use hedges as movement corridors. Today across swathes of lowland Britain they are often the only faintly decent bit of wildlife habitat left!

Enclosure of land by hedges hasn't been good news for all animals. The Great Bustard, a ground bird the size of a large goose, was reasonably common on extensive open grasslands and crop-growing areas in southern and eastern England in the Middle Ages. They simply could not have existed here when the wildwood left virtually no room for the large open, undisturbed areas this large ground bird relies on. Michael Shrubb (1934–2013), an ornithologist who spent a considerable amount of time investigating the impact of farming on birds, analysed a plethora of information which pointed to Great Bustards having arrived here in the late 15th or early 16th centuries.[vii] They were common in the lists of game presented at aristocratic feasts after this time but not before. Strong fliers that can migrate large distances, they would have been attracted over from continental Europe to the expanding open farmlands of southern Britain.

By the 1840s though, the bird was gone. Hunting for (and by) the aristocracy, the Inclosure Acts that partitioned up many open grassy and crop-growing areas with hedgerows, more efficient crop weeding, and perhaps more disturbance as the rural population grew in size, all sealed this magnificent but shy bird's fate. Only in recent years have they been re-introduced to one area of extensive downland in southern England.

Enclosure didn't just apply to the more intensively-managed open fields; it enclosed 'wastes' too. These areas of woodland, heath, rough grassland, scrub, marsh, and many a heathery slope of mountainside were the wildlife havens of pre-enclosure Britain. Slowly, much of this land was brought into cultivation, gradually depleting its shrubs, flowers, insects, mammals, and birds. It is from the late 18th century that biologists date a substantial decline in birds of prey, a general indication that what's not right at the top of the food chain signals that a great deal is wrong at the bottom!

For centuries, farmers had been draining the bits of wetland and wet grassland they could by hand-digging ditches to take the water away. What started slowly and piecemeal gathered pace by the 17th century as Dutch drainage techniques were introduced firstly into the then extensive fenlands (marshes) of East Anglia and, steadily, from there into much of lowland Britain (see Chapter 2). Maintained ever since as dry land by a complex system of huge drainage channels, banks, and powerful water pumps, the once wildlife cornucopia of the fens has become a major arable agricultural region in Britain for cereal and vegetable growing. It has elbowed out huge areas of natural wetland brimming with wildlife, an ecological disaster approaching that of the clearance of the wildwood.

For millennia, reedbeds, open water, tangles of wet scrub, and damp, rushy pasture had been commonplace, not just in extensive low-lying areas such as the fenlands, but also in small pieces of lowland here and there, often around rivers, ponds, and streams. In them a variety of fish thrived; so did a plethora of invertebrates including clouds of dragonflies and their smaller damselfly cousins; and ducks, stately herons, and elegant Common Cranes were also abundant. Great Bitterns would belch out their grunting, springtime calls and the reeds would have been alive with the songs of a multitude of small birds from plain brown Reed Warblers to straw and grey-coloured Bearded Reedlings. All of these and many more lost their place. A few of the marsh plants and insects survived in the drainage ditches that remained; an occasional otter or hunting harrier could grab a foothold too, but the bulk of Britain's phenomenal wetland wildlife drained away.

In Britain's hills and mountains (covering around one-third of the UK's land), regular burning of heather and similar heathy vegetation, combined with grazing by cattle on lower slopes and sheep higher up, came to alter great swathes of the vegetation beyond all recognition. Coarse grasses began to dominate the landscape except where deep peat clothed the rolling hills or where sheep were less able to penetrate. Vast tracts of our heathy uplands started to fade away, replaced with tough grasses that dominate large parts of our mountain slopes today.

TAKEN FLIGHT?

Farmland birds have fared extremely badly in recent decades. The British Trust for Ornithology (BTO) found that 19 farmland species (such as Tree Sparrow, Turtle Dove, and Grey Partridge) declined by 54% on average between 1970 and 2014.[xii] Some, like Turtle Doves – now confined to southern and eastern England and scarce even there – might be heading for extinction, their numbers having plummeted by 94% since 1994.

Although the rate of decline has slowed in the last decade, it remains a downward trend in spite of recent conservation measures being incorporated on a good number of farms. Research by Dario Massimino and colleagues found that farmland birds were doing better in western areas of the UK where they have been mostly stable compared to eastern areas where they have declined extensively, declines which have been attributed to long-term agricultural intensification, fragmentation of natural habitats, and urban expansion.[xvii] Even in the west, a farmland bird such as the Yellowhammer is declining as too much livestock grazing isn't leaving areas of more rank vegetation where the birds can find insects to feed their young, and because many hedges are losing their structure. Nor are these declines confined to lowland farmland. Twelve of 36 upland breeding bird species are red-listed as birds of conservation concern, species such as Whinchat, Ring Ouzel, and Curlew.

A report by DEFRA in 2015 states that 'many of the (bird) declines have been caused by land management changes and the intensification of farming that took place since the 1950s and 60s such as the loss of mixed farming, a move from spring to autumn sowing of arable crops, changes in grassland management (such as a switch from hay to silage production), increased pesticide and fertiliser use, and the removal of features such as hedgerows. The rate of these changes, which resulted in the loss of suitable nesting and feeding habitats, and a reduction in available food, was greatest during the late 1970s and early 1980s, the period during which many farmland bird populations declined most rapidly.'[xviii]

Britain is estimated to have lost around 44 million individual birds between 1966 and 2012 as a result of habitat destruction, much of it the result of a substantial increase in farm intensification.[xix]

Purple Bell Heather; insect-devouring Sundews rooted in Sphagnum mosses; Emperor Moths; Adders; and an abundance of small mammals, together with breeding birds such as Golden Plover, Whinchats, and Hen Harriers, would all have declined too. Take a walk today up to the highest sheep-grazed plateau in Snowdonia – the Carneddau – and the bleached grassy slopes dominated by toughies such as Mat Grass are home to an abundance of Meadow Pipits but few other birds except maybe some Carrion Crows, an occasional Raven and a few Skylarks on lower ground. Though the openness and relative solitude of this utterly treeless landscape does appeal to hillwalkers, like much of the British uplands, it is as Laurence Roche, Professor of Forestry at Bangor University between 1975 and 1992, described: one of the most degraded landscapes in Europe. Much of its wildlife has long gone.

In 2013, the UK's Joint Nature Conservation Committee (JNCC), the public body that advises the UK's governments on UK-wide and international nature conservation, reported that little over a fifth of our upland heathland and about half of our blanket bog (the extensive, gently undulating or flat peatbogs characteristic of much of our wet uplands) was in good condition.[viii] The rest was either slowly improving in condition, remained in poor condition, or was deteriorating further because of drainage, too much livestock grazing, and vegetation burning. Their next report is not due until 2019.

In Britain's uplands the transitional land between lowland and mountain (known as *ffridd* in Wales and *out-bye* in much of England) has proved particularly vulnerable. With herbicide sprays to kill bracken (and any other ferns) and equipment that could drain and plough the slopes, these hillsides – with patches of bracken, scattered trees, scrub, wet flushes replete with marsh orchids, and bits of semi-natural grassland – were especially at risk. Many were converted to grass pasture to support larger numbers of grazing sheep and attract yet more subsidies (then paid per head of livestock). In the 1980s, when capital grants from the EU's Common Agricultural Policy were at their peak, many farmers even got the bracken spraying done by helicopter! The temptation to convert the mix of wildlife habitats that echoed to the songs of Linnets, Tree Pipits, and Yellowhammers into vivid green grass of no wildlife value was, not surprisingly, huge for many an upland farmer. And it gave plenty of employment to local farm contractors with the necessary machinery. It was a bewitching agricultural enticement unmatched since in its bounty or damage to our uplands.

Scottish Natural Heritage (SNH) recorded a decline in heather moorland of about a quarter between the 1940s and 1980s, partly the result of planting it with conifers and partly because of burning, ploughing, and livestock grazing converting it to grassland.[ix] It is a trend that continued into the late 1990s.

Across the whole country, the most comprehensive information on habitat loss comes from a survey of Wales between 1979 and 1997 by the then Countryside Council for Wales (SNH's Welsh equivalent), although it was done over an extended length of time when further changes would have been taking place. It found that in the Welsh hills and uplands (land over 300 metres), at least 80% of the land consisted of semi-natural vegetation – heather moorland, coarse grassland, bracken, bogs, cliffs, and screes that might have been altered by people but not so drastically that it was effectively a man-made habitat.[x] The other 20% had been converted to ryegrass pasture or conifer plantations.

In the Welsh lowlands – four-fifths of the country – a very different picture emerged. Here, just 17% of the land consisted of semi-natural vegetation; it has probably declined further by today. An incredible 1.4 million hectares of the Welsh lowlands consisted of vegetation so modified it was of little or no wildlife value; one million hectares of it consisted of sown cultivated grassland, good for raising sheep and cattle yet exceptionally useless for wildlife save for a few Carrion Crows and maybe the occasional feeding Starling.

These million hectares are still very obviously green – rather bright paint-pot green in fact – a colour that convinces most passers-by that nothing has changed for centuries. However, the year-round brightness of the green indicates that what was there before – a flower-rich hay meadow or heather-clad heath maybe – has been ploughed up and the land sown with a mix of fast-growing ryegrass and clover. Either that or it's been changed more gradually through continuous grazing, by adding fertiliser, and perhaps using selective herbicides. Regardless, the end product is what farmers call 'improved grassland': grassland of virtually no wildlife value because it supports such a poor variety of wild plants and invertebrates. Semi-natural grassland is never such a vivid green; it contains a patchwork of green, olive, and straw-mottled hues, often much more straw-coloured than green through winter.

Unlike other flowering plants, grasses send out lots of side shoots when they are cut back. It's known as tillering and some grasses are better at it than others. It's why continuous livestock grazing suppresses most plants other than actively tillering grasses, and why poorly tillering grasses, as well as many meadow flowers, die out, since constantly removing their leaves kills them. It's also why a continuously mowed garden lawn becomes dominated by grasses; mow it less often and more flowers show up. Manure from grazing livestock also fertilises the soil, stimulating grasses and suppressing many non-grasses that prefer less fertile conditions.

A review by Dr Rob Fuller found that, between the 1930s and 1984, the area of lowland meadow (often rich in flowering plants and invertebrates) in England and Wales declined by a staggering 97%, and that the majority of remaining meadows, many historically cut for hay, had been degraded.[xi] Much of the loss was the indirect result of World War II (see Chapter 2) and the 'Dig for Victory' campaign which pushed farmers into ploughing up potentially flower-rich pastures and meadows and draining as much marginal land as possible for crop growing. Farming, for many years in the economic doldrums, became more profitable at the expense of farm wildlife. And the lost wetlands, scrub, wildflower meadows, and many other habitats never returned; by and large they have remained as more intensively-farmed croplands and pasture.

In the years since, even more farmland has been drained, ploughed up, and sown with cultivated grasses or altered equally drastically with fertilisers and more intensive grazing. Most former flower-rich hay meadows have become grass fields cut regularly for silage. Traditionally, hay crops are cut from meadows in summer when the crop is driest and the meadows then often grazed after the cut until the next spring. The loss of this annual regime has caused an enormous decline of typical meadow flowers – Oxeye Daisies, Cowslips, knapweeds, and much else – plus a huge range of invertebrates and small mammals. Birds such as Corncrakes, common throughout Britain until the 1960s, are now confined to meadowland in the Western Isles of Scotland.

The State of Nature report, published in 2016, concluded that, in relation to lowland farming:

> Looking at the long-term trends (1970–2013) of individual farmland species (over 1,300 different plants, invertebrates, mammals and birds), 52% declined and 48% increased. Among these, 34% showed strong or moderate declines, 36% showed little change, and 30% showed strong or moderate increases. Over the short term (2002–2013), the overall picture was unchanged.[xii]

According to the report, 40% of farmland invertebrate species have declined in numbers, about 25% have increased, and 35% show no change. The charity Butterfly Conservation has concluded that, overall, 76% of the UK's resident and regular migrant butterfly species have declined in either abundance or distribution (or both) over the past four decades while 47% of species increased on one or both measures.[xiii] They attribute a high proportion of the losses to agricultural intensification.

It's certainly not an objective measure of the decline of a wide range of invertebrates but motorists up until at least the 1950s in Britain regularly attached an insect screen to the radiator grill at the front of their cars in summer. Without it the dead insects would gradually block the radiator, substantially reducing its ability to keep the engine cool. Such a measure isn't necessary today – no one would contemplate it anyway!

Common bat species have actually increased – by 23% over the long term – in spite of substantial declines in invertebrates, their prey. But why? The Bat Conservation Trust's belief as it has been explained to me is that more stringent legislation protecting bat breeding and hibernating sites, a more positive public attitude towards these intriguing mammals, climate warming, and the wide variety of invertebrates they can feed on by foraging over large areas of land and water are all contributing factors.

The State of Nature report goes on to say:

> Our review of the factors driving changes to the UK's wildlife found that the intensive management of agricultural land had by far the largest negative impact on nature across all habitats and species. In one sense, it is no surprise that changes to our farmed environment have had more impact

than any other, simply because the habitat covers so much of the UK. Increases in agricultural productivity have been achieved through changes such as a switch from spring to autumn sowing of crops; the production of silage rather than hay in our pastoral farmland; and the increased use of chemicals over the long term. In addition, many marginal habitats, such as hedgerows and farm ponds, have been lost. Agricultural intensification affected nearly half of the species we studied and it was responsible for nearly a quarter of the total impact on our wildlife. Many factors have resulted in changes to the UK's wildlife over recent decades but policy-driven agricultural change was by far the most significant.[xii]

The report also documents losses of species and habitats in Britain's uplands. These are where the majority of the land is grazed, and where considerable areas of permanent grassland, heath, moorland, bracken, and scrub-dominated habitats on lower, gentler, and more accessible slopes have been obliterated and converted to intensively-managed, grass-dominated leys. What's more, these declines are continuing rather than reducing.

It says:

> Looking at the long-term trends of individual upland species, 55% declined (36% showed strong or moderate declines) and 45% increased between 1970 and 2013. Over the short term (2002–2013) the picture was similar; 54% of species declined and 46% increased.[xii]

No space left for wildlife: intensive cereal growing, Shropshire.

Dying from a nasty graze. Heavy sheep grazing on upland slopes in the Brecon Beacons

The report also draws on the results of the PREDICTS project, a collaboration funded by the Natural Environment Research Council (NERC) and carried out at a number of institutions. It estimates the average abundance of originally present species – nearly 40,000 plants, fungi, and animals – relative to their abundance in undisturbed habitat to give a measure of how 'intact' our plant and animal communities and their habitats are. The UK came 189th out of 218 countries! The only reasonable 'intact' areas in the UK are mostly the more remote uplands and mountains in Wales, Scotland, and Northern England, illustrating rather dramatically the enormous extent to which we have messed up much of the UK's wildlife habitats and the species associated with them. Agriculture, urban encroachment, planting conifers; all have taken their toll. But with farming occupying almost three-quarters of our land, the frequent claim of farming representatives such as the National Farmers' Union (NFU) that 'farmers are the custodians of our countryside' has a distinctly hollow ring.

Losses of farmland plants have been substantial too. According to Plantlife, the UK wild plant conservation charity:

> In the uplands, traditionally managed hay meadows are now few and far between. As you travel through National Parks and Areas of Outstanding Natural Beauty such as the Lake District, the Pennines and the Yorkshire Dales, the impressive physical landscape of dry stone walls is still there but much of the botanical richness is gone. On arable farmland, management

intensification has probably gone further than anywhere else. Once familiar cornfield flowers are now of conservation concern with species such as Corn Buttercup and Red Hemp-nettle going from "weed" to "rarity" in a few decades. Cornflower numbers are down by around 99% in 50 years and today are practically extinct as wild flowers.[xiv]

Seven of the UK's former arable plants – flowers associated particularly with lowland cereal growing and other cultivated land – are extinct; another five are critically endangered (including Corn Buttercup) while 34 more are considered to be endangered or vulnerable. They include the canary-yellow Corn Marigold, which was considered a serious weed in Victorian times, and the delicate, pink-blue Grasspoly, which is now restricted to a handful of locations. Developments within arable farming since the 1960s have caused great changes to this flora, with many species unable to adapt or survive the revolution in farming methods. Advanced seed cleaning to remove any unwanted seeds, increased use of fertilisers, new high-yielding farm crop varieties, and the introduction of selective herbicides have all resulted in more efficient control of such undesirables in farmers' crops. The shift from spring to autumn cultivation and a general reduction in crop diversity have further contributed to their decline. So-called 'arable weeds' are the most threatened group of wild plants in the UK today.

There is, though, some evidence of more recent overall improvement. The State of Nature report concludes that while the distribution of farmland flowering plants declined by an average of 7% over the long term (1970–2013), that rate of decline has slowed considerably in more recent years and has reversed to give a 2% increase between 2002 and 2013.[xii]

Farming has implications not only for wildlife on farms but also for habitats affected by the pollutants it produces. Soil, fertilisers, and pesticides can run into watercourses after heavy rain or be deposited on sometimes distant off-farm habitats by wind. Farming makes a sizeable contribution to airborne pollution in the UK (see Chapters 8 and 12). In 2014, it accounted for almost 83% of all ammonia emissions in the UK, over half of it from cattle and pig manure.[xv] However, while these emissions fell by 24% between 1990 and 2014 due to fewer cattle and more efficient fertiliser use, they started rising again between 2013 and 2014.

When ammonia falls as rain after being taken up into the atmosphere, it can change the composition of the vegetation it falls on. This encourages rank grasses to grow at the expense of plant species that naturally require less nutrient-rich conditions – heathers and heaths for instance. It can also contribute to the acidification of upland streams, changing their ecology too.[xvi]

Wildlife losses are not simply desperately bad news for the species themselves or for those of us that appreciate natural history. They also reduce the excitement and pleasure many families used to derive from encountering some of these plants and animals once common in our farmed countryside. Furthermore, there is growing concern that their decline and loss is going to have an impact on the ability of whole ecosystems to continue functioning. We – us humans – are an integral part of those ecosystems. For instance, a recent study examined declines in over 4,000 plants and animals over four decades and found 'significant net declines among animal species that provide pollination and pest control'.[xx]

So what does this suggest? Firstly, bees are highly effective pollinators; hence the widespread concern about their decline and the possible role of pesticides in causing it. A study assessed the implications of the decline in nectar- and pollen-producing plants associated with hay meadows. It found that of the 97 food plants that bumblebees prefer, 76% have declined over the past 80 years because hay meadows have been ploughed up.[xxi] Other groups of insects – such as hoverflies, moths, and butterflies – are also important pollinators. It doesn't end there though; of secondary importance are many others, including soldier beetles, glow-worms, and wasps.

Farm crop pollination relies on these free and unsung workers of the agricultural industry, yet we seem to largely ignore the vital task they perform. Without them, our food supplies would dwindle frighteningly rapidly. Other species act as free pest controllers and we ignore them at our peril too. Carabid beetles, ladybirds, spiders, centipedes, wasps, dragonflies, and damselflies, harvestmen – I make no apology for the length of this list – hoverflies, soldier beetles, glowworms, some ants, plus numerous insect-eating birds and mammals, all are predators that act as natural enemies of crop pests. We deplete them at our peril.

Looking down over the patchwork fields of parts of lowland Britain from an airliner flying above, the tablecloth below of paint-pot green pasture, cream cornfields, and vibrant yellow rapeseed dissected by meandering hedgerows and dotted with attractive copses of trees probably reassures most viewers that our bucolic, rural haven is as idyllic as ever it's been. But it isn't. The vibrant colours are actually a giveaway. With crops fertilised to gain optimal productivity, any weeds and pests (plus many beneficial species into the bargain) reduced or eliminated with pesticides, and with fertilised ryegrass pastures supporting intensive livestock grazing, only the surviving hedgerows and copses of trees retain any significant wildlife. Other extensive areas of our lowlands don't even have many hedges left! Our lowland landscape has become an industrially farmed workplace which is unwelcoming for most of our wild plants and animals. Ironically, the intensively-managed fields of these agri-businesses would be far better for wildlife if they were converted to housing; the resulting gardens with their shrubby borders, scatter of small trees, and the occasional pond would support much more wildlife than these artificially-coloured fields ever will.

On the higher reaches of our hills, moors, and uplands, the changes wrought by modern agriculture have been less severe. Large swathes of good wildlife habitat remain though much else has been altered by centuries of livestock grazing – much of it too intensive – and vegetation burning, slowly but surely reducing the diversity of plants and animals that were once its inhabitants too. Over the last five decades in particular, sheep have been stocked too heavily (encouraged for many years by subsidies paid per head), grazing away a good deal of the plant diversity our hills formerly nurtured, depleting in turn our upland wildlife more widely.

We cannot turn back the clock, at least not on any significant scale. We cannot recreate the wildwood. A few of our remotest places, parts of Scotland or Central Wales maybe, could be left entirely to nature without any land use save for walking and hiking – so-called 'rewilding' (see Chapter 13). These areas could be allowed to revert to moor and bog, scattered scrub, and woodland, replete with a wide range of plants and animals that would re-colonise naturally or be bolstered by re-introductions. Many of the wildlife habitats that we cherish today – lowland heaths, flower-rich limestone grasslands and meadows, blanket bogs, and heather moors in our uplands – rely on some form of farming, light sheep or cattle grazing, or hay making, because the wildlife riches of these habitats have evolved over millennia with and because of their farming use. There can be no doubt that we need the food that our farmers provide but what's needed even more is a better balance between this requirement to provide healthy food and a wildlife-rich countryside, a balance that is currently way out of kilter.

Endnotes
i   'Small farmland bird declines, gamebird releases, and changes in seed sources,' by Alan Larkman et al. *Wildlife Conservation on Farmland, Volume 2*. Edited by David Macdonald and Ruth Feber. Oxford University Press, 2015.
ii   *Silent Fields: The Long Decline of a Nation's Wildlife*, by Roger Lovegrove. Oxford University Press, 2007.
iii *Hedges*, by E. Pollard, M. D. Hooper, and N. W. Moore. Collins, 1974.
iv   *Trees and Woodland in the British Landscape*, by Oliver Rackham. J.M Dent and Sons Ltd, 1976.
v   'Hedgerows: a guide to wildlife and management,' People's Trust for Endangered Species, 2014.
vi   http://www.hedgerows.co.uk/Introduction-%20What%20we%20do.htm.
vii *Birds, Scythes and Combines: A History of Birds and Agricultural Change*, by Michael Shrubb. Cambridge University Press, 2003.
viii '3rd UK Habitats Directive Reporting 2013,' Joint Nature Conservation Committee, 2014.
ix   Scottish Natural Heritage TREND NOTE, number 008, Land Cover Change 1940s – 1980s: Heather moorland. 2003.
x   *Habitats of Wales*, Edited by Tim Blackstock et al. University of Wales Press, 2010.
xi   'The changing extent and conservation interest of lowland grasslands in England and Wales: A review of grassland surveys 1930-1984,' Rob Fuller, 1987. *Biological Conservation*, 40: 281-300.
xii 'State of Nature 2016,' D.B.Hayhow et al. The State of Nature Partnership, 2016.
xiii 'The State of the UK's Butterflies 2015,' Butterfly Conservation.

xiv England Farmland Report: 'And On That Farm He Had,' Plantlife, 2013, and 'New Priorities for Arable Land Conservation,' Plantlife, 2007.

xv   Agriculture in the UK 2015. Defra and devolved Governments, 2016.

xvi Observatory monitoring framework: indicator data sheet DC1: ammonia emissions, Defra, 2015.

xvii 'Multi-species spatially-explicit indicators reveal spatially structured trends in bird communities,' Dario Massimino et al, 2015. *Ecological Indicators* 58: 277-285.

xviii 'Wild Bird Populations in the UK, 1970 to 2014,' Defra, 2015.

xix 'Birds of Conservation Concern 4: the population status of birds in the UK, Channel Islands and Isle of Man,' Mark Eaton et al. *British Birds*, 108: 708-746.

xx   'Declining resilience of ecosystem functions under biodiversity loss,' Tom Oliver et al, 2015. *Journal name:Nature Communications* 6; article no. 10122.

xxi 'Declines in forage availability for bumblebees at a national scale,' Claire Carvell et al, 2006. *Biological Conservation*, 132: 481-489.

CHAPTER 4

TAKING OFF THE CAP

*In any bureaucracy, the people devoted to the benefit of the bureaucracy itself
always get in control, and those dedicated to the goals the bureaucracy is supposed
to accomplish have less and less influence, and sometimes are eliminated entirely.*

JERRY POURNELLE, AMERICAN SCIENCE-FICTION WRITER, ESSAYIST, AND JOURNALIST

For four decades farm advisors have hung on every word of it. Veritable armies
of government officials in 28 states have drawn up copious rules to implement it.
Conservation organisations and other rural campaigners have fretted long and
hard over its every phrase. Finance ministers have frequently struggled to justify
its enormous cost. And the EU's 14 million farmers have been the beneficiaries
of the EU's Common Agricultural Policy (CAP), designed to keep them farming
and producing much of our food.

In the UK all that is set to change, perhaps drastically. The vote on 23 June
2016 to leave the EU will set our farmers free from the CAP for the first time in
over 40 years, a momentous change that many farmers voted for. Yet to consider
what support structures, if any, should replace the CAP, and how our farmland
can be made considerably more wildlife friendly than it currently is, requires an
understanding of how the CAP has brought UK farming to where it is today.

Created over half a century back, the CAP – much modified since its inception
in 1962 – is still the overarching mechanism governing farming Europe-wide
from the Arctic to the Mediterranean. In its time it has consumed as much as
70% of the EU budget (today it is under 40%); in 2015, it cost €59 billion. It
has been roundly criticised for the environmental damage it has stimulated, its
contribution to inflated food prices in Europe, and for stalling development in

poorer countries by preventing many from exporting food to the EU. The UK eventually joined up by becoming part of the EU in 1973.

According to the official website of the European Union, the main aims of the CAP are: to improve agricultural productivity so that European consumers have a stable supply of affordable food, and to ensure that EU farmers can make a reasonable living.[i] It also ensures the safety of our food, shields farmers from excessive price volatility and market crises, helps them invest in modernising their farms, sustains viable rural communities, creates and maintains jobs in the food industry, protects the environment, and sets standards for animal welfare. Many of its critics, though, would take issue with much of that list!

WHAT THE EU's FARMERS PRODUCE EACH YEAR

Cereals:	300 million tonnes
Sugar:	17 million tonnes
Oilseeds:	20 million tonnes
Olive oil:	2 million tonnes
Apples:	10 million tonnes
Pears:	2 million tonnes
Citrus fruit:	11 million tonnes
Wine:	160 million hectolitres
Beef and veal:	8 million tonnes
Pig meat:	20 million tonnes
Poultry meat:	12 million tonnes
Eggs:	7 million tonnes
Milk:	150 million tonnes

(Source: EU data, 2015.)

Its origins are in the late 1950s after the founding members of the European Community emerged from over a decade of severe food shortages during and after World War II. The first CAP was adopted by the six founding member states, coming into force in 1962 and slowly 'harmonising' different member states' supports to 'create a level playing field' and allow freedom of trade across the developing EU. In the 1970s, attention began to focus on initiatives such as modernising farms, promoting professional training, and re-invigorating the agricultural workforce by encouraging older farmers to take early retirement. In 1975, steps were taken to assist farmers working in difficult conditions, such as hill farmers, while also reducing the burgeoning CAP budget.

Although the CAP was successful in meeting its objective of moving the EU towards self-sufficiency, by the 1980s the over-generous subsidies and guaranteed prices for most farm products had created near permanent surpluses of the major farm commodities – the infamous 'wine lakes and butter mountains' – some of which were exported and sold off cheaply, stored at huge cost, or disposed of within the EU. The CAP was rapidly falling into disrepute, made even worse by the realisation that the subsidies and capital grants paid to many farmers were fuelling an unprecedented destruction of wildlife habitats (see Chapter 3).

Reforms in the 1990s and early 2000s split CAP funds into two: known as 'pillars' in EU jargon (and there is plenty of EU jargon). Pillar 1 consists of direct payments – subsidies now known as the Basic Payment Scheme (BPS) – made to all farmers with at least five hectares of land. Across the UK in 2015, it was worth £3.1 billion. Since the mid-2000s these subsidy payments have not been linked to the quantity of farmed produce but calculated instead per hectare of farmed land.

At a stroke this removed a huge incentive for farmers in the UK's hills and uplands who had destroyed wildlife habitat such as heather moor and converted it to pasture sown with cultivated grasses in order to support yet more livestock. More livestock meant an increased subsidy since it was initially headage linked. Largely because of the change from headage payments to payments per hectare, the number of sheep and lambs in Wales fell from 11.2 million in 2000 to 9.7 million in 2014.[ii] The impact of less grazing is particularly apparent in small parts of our uplands where vegetation is becoming more rank and heathy plants like Bilberry are returning along with scattered trees.

Since 2003, Pillar 2 – known as the Rural Development pillar – has included money to fund agri-environment schemes which farmers can join voluntarily. These reward farmers for implementing more wildlife-friendly farming methods or recreating habitat, albeit after much of the Pillar 1 subsidies fuelled so much wildlife habitat destruction! The 2003 reforms also introduced something called 'setaside', a policy of taking farmland out of production in order to reduce the height of the so-called wheat mountains while delivering environmental benefits. The wheat mountains gradually subsided and the wine lakes became ponds. Welcomed by conservation bodies, setaside recreated some much needed wildlife habitat. But it didn't last. With shortages of some crops, cereals in particular, over subsequent years, setaside was abolished in 2008.

The 2003 reforms also brought in 'cross-compliance', yet another piece of EU jargon in which the 'cross' is superfluous. It means simply that farmers receiving their subsidies have to comply with a set of minimum environmental, animal welfare, and safety standards. It's a sound principle; in reality though, the standards are undemanding and provide little in the way of significant benefits for the environment or wildlife. Flagrantly failing to comply can result in some farmers having their subsidy payments reduced but sanctions are rarely invoked.

Compliance requires farmers to keep land in good agricultural and environmental condition, prevent soil erosion, retain its organic matter and structure, avoid the deterioration of habitats, prevent water pollution, and retain certain landscape features such as hedges, stone walls, ditches, and natural ponds. It seems to me that much of this should be regarded as good farming practice. However, compliance also includes some rather contradictory and anti-wildlife measures too. One is the need to 'avoid the encroachment of unwanted vegetation on agricultural land'. What this means is that farmers don't get their subsidy on land that can't be farmed, 'invaded' by scrub or bracken perhaps. So they can cut trees, gorse, and other scrub or burn it (provided they do it before mid-April in the uplands) and use the cleared land to graze livestock, thereby destroying that habitat. It's absurd.

Criticism of the CAP has centred on the encouragement it has given to destroying wildlife habitats – especially in the past with production-based subsidies – and because farmers with larger farms (who need subsidising the least) are the biggest beneficiaries, getting the greatest amount of support. In fact, 80% of the total farm subsidy is paid to 20% of farmers! A small number of UK farms and their already-rich landowners receive over £1 million a year. According to the then Scottish Secretary for Rural Affairs and the Environment, Richard Lochhead, the top five recipients in Scotland (large estates used for farming and grouse shooting) have received between them over £7.5 million annually.[iii]

Effectively setting the agenda for the management of half of the EU's countryside (over 70% of our land in the UK because we have far less forest than most EU countries (see Chapter 1)), the European Commission makes this statement about the value of farming:

> Farming has contributed over the centuries to creating and maintaining a variety of valuable semi-natural habitats. Today, these shape the many landscapes throughout the EU and are home to a rich variety of wildlife. Farming and nature influence each other. Thanks to the successive reforms of the CAP, our farming methods are becoming more environmentally friendly.
>
> Today's farmers therefore have two roles – producing our food and managing the countryside. In the second of these they provide public goods. The whole of society – present and future – benefits from a countryside that is carefully managed and well looked after. It is only fair that farmers are rewarded by the CAP for providing us with this valuable public good.
>
> Protecting biodiversity and wildlife habitats, managing water resources and dealing with climate change are other priorities that farmers are required to respect.[iv]

The statement conveniently airbrushes over the enormous damage and degradation to wildlife habitats and species caused by CAP-funded policies particularly between the 1970s and the 1990s. Reforms to the CAP to better balance the twin requirements of food production and wildlife conservation have been lethargic, belatedly trying to put things right after the damage has been done. With its policies needing to be thrashed out and agreed by 28 member states that have farming industries ranging from alpine cheese-making to wine production and from sheep grazing to rice growing, it's a miracle that any composite policy can be agreed at all. Freed from the stranglehold of the CAP, the UK Government and our devolved administrations will be able to devise their own farm policies better suited to each country's needs, providing an unprecedented opportunity for our four governments (agriculture is devolved) to cater far better for wildlife and support farming at the same time.

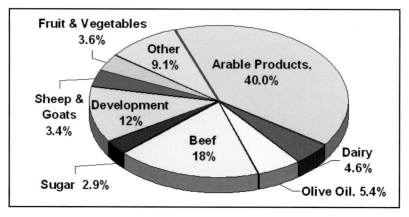

Farm Spending by Sector (Source: EU data, 2015.)

Each UK country administers its BPS subsidy payments separately using its own agency. In England it's the work of the Rural Payments Agency (part of Defra), and its scheme rules run to a redoubtable 111 pages! The rates paid per hectare for 2016 were €175.27 on lowland farmland, €174.01 on upland, and €45.97 on moorland.[v] Maps define which zone(s) a farmer is in. These payment levels mean that a lowland farmer in England with 200 hectares of farmed land would expect to receive €35,054 per annum in BPS subsidies. A hill farmer with 40 hectares of upland and 250 hectares of moor (used only for livestock grazing) would receive €18,453. To encourage young people into farming, BPS payments can be topped up by a quarter for the first five years.

Wales and Northern Ireland have a mix of rates at present (a hangover from a past set of different rates for different land) but by 2019 they will be paying a standard €120.80 and €229 respectively per hectare on all farmland, making no distinction between lowland, upland, and moor.[vi] It means that come 2019 hill farmers in both countries would be substantially better supported than those in England (especially if they farm in the relatively small area of upland in Northern Ireland) if subsidy payments were not to change when we leave the EU.

Further illustrating how greatly subsidies vary across the UK, Scotland's rates by 2019 are expected to be €145 per hectare for lowland (about 1.8 million hectares), just €25 per hectare for 'better quality' rough grazing land (about one

Increasingly intensive. Spraying glyphosate to kill weeds before the sown crop emerges.

million hectares), and only €7 per hectare for the high uplands which have the poorest rough grazing in the country (about two million hectares).[vii]

The BPS continues the practice, in place since 2003, of making payments based on farm area and not on livestock numbers or the amount of crop produced. But there are exceptions to this and Scotland – alone in the UK – negotiated hard to be able to add to its lowest BPS payment (€7 per hectare) an additional €25 per sheep on its poorest grazing land.[vi] Although the Scottish Government's Department of Rural Affairs, Food and Environment points out that their intention is to protect its sheep-meat industry and not to encourage an increase in stocking rates (more ewes equals more payments), it is nevertheless likely to be an incentive to graze more sheep on the Scottish hills and moors, some of the most wildlife-rich habitat in Scotland. Scotland is also re-introducing a payment per head for some beef cattle including increased supports for cattle grazing in the islands. Such payments are aimed at topping up comparatively low BPS payments but the overall policy of paying for production rather than solely for land area is almost always bad news for wildlife.

A significant part of the thinking behind these Scottish decisions is the fear by politicians and some farming interests of what is termed 'land abandonment'. It suggests that there is some vital need to retain hill farming whatever its cost, however difficult the terrain, and however low the productivity in terms of livestock numbers. But why? Leaving mountain and hill land un-grazed to slowly redevelop its more natural vegetation and attract a much wider variety of wildlife is a perfectly valid alternative. Better-developed vegetation on hill slopes can also play a significant role in absorbing more rainfall than closely-grazed grassy swards ever can (from which rain runoff can be notoriously rapid), potentially reducing the risk of flooding rivers downstream. It doesn't bode well for post-CAP years, when Scotland will be solely responsible for its own farm policy.

With the area-based BPS system, the larger the farm, the greater the subsidy it attracts. However, in recent years the EU has agreed that member states can apply a limit to this very generous part of the CAP, thereby addressing the long-held criticism that too much subsidy goes to the largest farms with the best land who therefore need the support the least. Wales has introduced a sliding scale and a limit of €300,000 per farm but Scotland has set its cap twice as high at €600,000. These are still huge amounts of public money paid, in the main, to farms making good profits from the crops they grow or livestock they raise. They are eye-watering levels of support when other, often struggling, rural businesses such as village shops, post offices, and similar community facilities get no such annual supports to keep them in business.

English farmers, though, are not capped at all, suffering merely a 5% reduction in their per hectare payment on claims over €150,000 annually. With some of the largest lowland farms and estates in the UK, a number of farmers and farming estate owners will be attracting annual payments reaching €1 million per annum; some of the biggest crop-growing companies farming large areas of prime English lowland will be claiming far more! So a great deal of CAP money does indeed still go to the major, well-off farmers who need it least. This is especially true in England, where large landowners and farming companies retain significant influence over successive UK governments and are literally milking the system. Once we have left the EU, and the amounts of funding paid to farmers is in direct competition with other calls on expenditure such as the NHS, social supports, and education, the UK Government and the devolved administrations are likely to be under much closer public scrutiny.

This unnecessary largesse also takes away cash that could otherwise be devoted to rolling out more agri-environment schemes on farms in order to support wildlife conservation. Dame Helen Ghosh, Director General of the National Trust, one of the UK's largest landowners, said in August 2016 at the Countryfile Live Conference:

> I should of course say here that the National Trust has been a significant beneficiary of these subsidies. Our tenant farmers are recipients in their own right, and the Trust itself receives about £11 million a year, £3 million in direct subsidy and £8 million for Environmental Stewardship [agri-environment] schemes. But we spend all of the money we get on conservation – and indeed accept that in the future the amount of support we get might well fall.

In an attempt to make the BPS payments a bit more environmentally and wildlife friendly, since 2015 all farmers receiving it have had an obligation to undertake agricultural practices that are beneficial for the climate and the environment. But these attract payments *additional* to the BPS! Referred to as 'greening', in England in 2016 they added an incredible €77.71 per hectare on lowland, €77.15 per hectare on upland, and €20.39 per hectare on moorland to the BPS payments. What farmers are obliged to do in order to get these greening payments is, in reality, not a lot. They are a dog's breakfast of measures that do little to improve the lot of wildlife on any farm, although it's important to understand their presumed benefits in order to assess whether they should be continued post CAP.

Three elements make up the greening requirements: crop diversification, Ecological Focus Areas (EFA), and permanent grassland. The rules for implementing them are fiendishly complicated. For crop diversification, farmers with ten hectares or more of arable land must grow at least two different crops with the main crop not covering more than 75% of the arable area. Farmers with more than 30 hectares of such land must grow at least three different crops with the two main crops not covering more than 95% of the land. The choice of crops is huge – the list runs to ten pages! Farmers with more than 15 hectares of arable land must also adopt the EFA greening rules but those with less than ten hectares of arable land don't need to do anything differently; seemingly they meet the crop diversification and EFA rules automatically. How that makes sense if a farmer with less than ten hectares of arable land can plant it all with one crop, sugar beet or wheat for instance, is difficult to fathom. Moreover, farms that have 75% of

Devoid of wildlife: a distant scatter of small trees in an otherwise unbroken expanse of crops, Dorset.

Tens of millions of sheep (these two are Swaledales) graze most of Britain's uplands.

their land in grass pasture don't have to implement these measures either. There are other exemptions too, all of which provide farmers trying to understand the rules with a headache and bureaucrats across Europe with work.

So what wildlife value does arable crop diversification provide? If all such crops, whether they are peas, barley, turnips, or linseed, are grown intensively using selective herbicides and insecticides, it doesn't make any measurable difference what crop is being grown because its wildlife value will be extremely low. But research suggests that some open land birds do better where there is more crop variety; Skylarks in particular seem to be more abundant and have somewhat greater breeding success in such circumstances.[viii] That's a small plus. Peas and beans have proved to be popular crops since these measures were introduced, nearly doubling in area to 230,000 hectares in 2016 compared with just two years earlier.[ix] Neither of these crops is insect pollinated so they provide no wildlife gain.

Farmers with 15 hectares or more arable land also must have EFAs occupying at least 5% of it. EFAs can be strips of uncultivated land at the sides of arable crops or within them; short rotation energy crops of fast-growing trees; ditches and hedges on the farm; small areas of planted woodland; catch crops of quick-growing plants (some of which can attract insects) used as a soil cover between main crops to prevent nutrient runoff; traditional stone walls; land areas left fallow for short periods, and others. Oddly, there is no stated obligation for the farmer to maintain hedges and stone walls; they simply have to exist to qualify, and most farmers will obviously have one or other!

'The lack of any management requirement to maintain features such as ditches and hedges so that they survive long term is a major flaw in the EFA idea,' comments Tom Lancaster, Senior Land Use Policy Officer at the RSPB. Of the ten different EFA measures that can be implemented, the UK is implementing just half of them. Scotland, for instance, is implementing buffer strips, fallow areas, hedges and ditches, catch and cover crops, and nitrogen-fixing crops, in many ways the easiest to implement that cause farmers the least effort.

Scepticism over whether EFAs will deliver much additional and practical wildlife benefit has grown. Measures such as planting catch crops will help reduce the constant problem of soil and fertiliser runoff on lowland farms, thereby reducing pollution and sediment flow into streams and ditches, indirectly benefitting aquatic species in those habitats. However, the extent to which the options either lead to a change in farm management or, alternatively, simply replicate activities that would have taken place even without the greening payments is a critical issue. There is a very high likelihood that many farmers will be able to meet the requirements without changing what they do now![x] Many farmers had opted for growing nitrogen fixing crops to satisfy their EFA requirements.[ix] But in 2017 the EU introduced a ban on any pesticide use on such crops so more farmers are now turning to alternatives, such as hedges, buffer strips and field strip options, thereby providing some small wildlife gains.

The permanent grassland provision comes into play only if the percentage of permanent grassland in each UK country – relative to the area of all agricultural land – falls by more than 5% from a baseline established in 2015. Should that happen, farmers who have ploughed up permanent grassland may have to re-instate it, and restrictions would be introduced to prevent further losses. In practice, such a fall is more likely in England than in Wales or Scotland because they have far less agricultural land devoted to arable crops than in the English lowlands. Member states are also required to designate permanent grasslands which are environmentally sensitive and need strict protection in order to conserve their flora. But any permanent grassland of significant wildlife value in the UK is already likely to be protected against being ploughed up or having synthetic fertilisers and pesticides applied to it because it will have been designated as a Site of Special Scientific Interest (SSSI) to reflect its national or international importance. A farmer with such a site would need permission from the relevant wildlife authority such as Natural Resources Wales if they wanted to alter its use.

There are, exceptionally, permanent grasslands where occasional ploughing is traditional and necessary for maintenance as important wildlife habitat. An example is machair, a low-lying grassy plain on very sandy, alkaline soils found on parts of the north-west coastline of Scotland (and Ireland), in particular in the Outer Hebrides. Often carpeted with flowers including orchids, Yellow Rattle, Cowslips, and many others, machair provides a special habitat for lots of invertebrates and breeding birds including Corncrakes and Dunlin. Scotland has rightly exempted machair from the permanent grassland restrictions.

Alan Matthews, Professor Emeritus of European Agricultural Policy in the Department of Economics at Trinity College, Dublin, Ireland, has summarised the value of greening like this:

> Even though 30 per cent of each Member State's direct payments is now allocated to greening, very little additional benefit for the environment will be obtained. In the case of EFAs, for example, a large number of farms are exempt, including all grassland farms as well as smaller arable farms.
>
> More important is that the elements permitted to count towards an arable farm's EFA have been expanded to include not just fallow land, buffer strips and landscape elements such as trees, hedges, ponds and ditches but also areas sown to legume crops or catch crops. As many farms across Europe already meet these requirements as part of their normal farming practice, the changes are likely to affect less than 1 per cent of the arable area.
>
> However, the biggest problem with EFAs is that farmers, rightly, see the obligation as just another set of bureaucratic hoops to jump through in order to receive the direct payment. There is no link made to encourage active management of the land to maximise the potential for biodiversity, nor to make use of farmers' knowledge and experience as to what might be the most appropriate measures to take.
>
> By trying to encourage more sustainable land management through the greening payment, the CAP reform missed the opportunity to put more money into agri-environment schemes which can be targeted to take account of different needs in different areas and in which farmers can see clearly that they are being paid to supply a valuable public good.[ix]

What's worse than the inordinate amounts of extra bureaucracy the greening requirements generate is that they account for a third of Pillar 1 CAP spending. This is about €14 billion annually EU-wide; an incredible 9% of the total EU budget, all for very little environmental gain. It's a scandalous waste of taxpayers' money.

As Professor Matthews makes clear, where the CAP does provide a meaningful contribution to more wildlife friendly farming is by funding agri-environment schemes (now referred to as agri-environment climate schemes because they have to contain measures to address farming's contribution to climate warming). These are funded from Pillar 2 of the CAP. In 2015, the UK had a budget of €3.1 billion for the Pillar 1 BPS subsidy payments but only a fifth of that for agri-environment payments to help restore farmland wildlife. Post CAP, with no EU restriction on how much and on what any farm budget should be spent, the UK's governments could decide to reduce the subsidy payments and boost agri-environment cash. They could equally determine to remove subsidies entirely and link all payments made to farmers via agri-environment schemes. Under the CAP, these schemes are voluntary for the farmer; should they remain voluntary or is our farmland wildlife in such a precarious state that they should be made compulsory?

Each UK country runs its own agri-environment scheme, and the differences between them reflect the varying farming circumstances and wildlife habitats across the UK. England's scheme, Countryside Stewardship (CS), run by Natural England on behalf of Defra, aims to improve the wildlife value of around 2.5 million hectares of farmland. It includes measures to protect soils; reduce flooding and erosion; protect landscapes, wildlife, and historic features; and improve public access and education. Any farmer entering it commits himself for a minimum period of five years to adopting environmentally-sympathetic measures on at least part of his farm. In return, he receives payments that reflect the extra costs involved (for additional fencing; for sowing a particular crop; or for creating flower-rich strips along arable fields for example) and the reduction in crop or livestock productivity by applying these measures, so-called income foregone. So they might fund planting trees, hedges, and copses; digging ponds; re-creating a hay meadow redolent with wild flowers and grasses; leaving a wide unsprayed and un-fertilised grassy strip around the outside of an arable field; fencing off streamsides to keep livestock away, which allows streamside vegetation to flourish; for retaining scrub around drainage ditches; for repairing dilapidated stone walls; and much more. Each implemented measure has a calculated annual cash value – agreed at the moment by the EU but, in England, presumably by Defra post-CAP – paid to the farmer in addition to BPS payments.

Creating scattered bits of wildlife habitat on a farm – a small copse of trees, a pond, some plant-rich field margins perhaps – surrounded by intensively-managed farmland has a benefit for wildlife. Yet it's far more beneficial if these habitats can be connected up within a farm and between farms, thereby encouraging the movement of less mobile species. An isolated bank of flowers and grasses will rely on seeds blowing in on the breeze or carried by animals to become richer in a more diverse range of plant species; likewise it might be only flying insects that manage to reach it. Joining up habitats from farm to farm allows more species movement, an important factor since whole populations of species need to move (some already are) in response to a warming climate. As they are voluntary, agri-environment climate schemes don't encourage a joined up approach; that's why The Nidderdale Landscape Partnership (Chapter 9) has endeavoured to get as many farmers as possible in this large upland valley in the Yorkshire Dales to participate in England's agri-environment climate scheme.

Glastir is the Welsh Government's current agri-environment climate scheme run by its Department for Environment and Rural Affairs. It is directed at managing soils to help conserve carbon stocks and reduce soil erosion; improving water quality and reducing surface run-off; managing water to help reduce flood risks; enhancing wildlife and biodiversity; managing and protecting landscapes and the historic environment; and creating new opportunities to improve public access and understanding of the countryside. Agreements again run for five years and by July 2016, 58% of Wales' agricultural land was in Glastir.

The scheme has a number of elements including Glastir Entry, the most basic element, in which farmers have to accept a whole farm code including measures such as retaining all in-field trees, not ploughing close to watercourses, not burning vegetation in rocky areas, and many others. Each measure proposed by a farmer is scored and an applicant has to achieve a threshold score to enter their farm. By July 2016, 561,000 hectares of farmland had been accepted in Glastir Entry.

However, the Welsh Government is investing more of its agri-environment climate budget into the Glastir Advanced element. It's aimed at delivering environmental improvements to benefit wildlife habitats, species, soil, and water in areas of particular importance for wildlife, including protected sites such as SSSIs. It's a competitive scheme; applicants have their farm proposals scored for the wildlife habitats and species present or proposed for restoration, for protecting water quality, restoring peatlands, and several other measures. By July 2016, over 384,000 hectares of farmland had been entered into Glastir Advanced.

Another element of the scheme, Glastir Commons, is aimed at reducing the impact of over-grazing on common land by paying for smaller livestock numbers or for less grazing time each season; it already covers over 70% of common land in Wales, 113,000 hectares by July 2016. Glastir Woodland is specifically targeted at funding new areas of farm woodland planting and improving the management of existing woodlands, some of which desperately need livestock fenced out so that young trees can regenerate to guarantee woodland survival into the future. Tree planting helps meet climate change criteria by absorbing carbon dioxide; it also provides valuable wildlife habitat.

All of these schemes had forerunners and many farmers still have agreements on the conditions they signed up to in these earlier schemes. Since 2005 in England, well over 50,000 farmers and other land managers have been signed up to Environmental Stewardship, the earlier version of Countryside Stewardship. These early agreements cover over 6.4 million hectares or 70% of farmland in England. But the vast majority of these are 'entry level' schemes that provide very limited wildlife gains. 'Entry level' has now been discontinued because it's a poor use of resources; instead, applicants for CS will be expected to adopt some habitat creation or better management of existing habitat at their own expense over and above what little is delivered by cross-compliance or greening before they even apply to join. Like Glastir Advanced it will be more focused than previous agri-environment schemes have been on delivering local environmental priorities. Payments will be offered to those who propose to make the best environmental improvements within their local area. The scheme is competitive and targeted with most money going to those farmers delivering the most wildlife and environmental benefit.

The CS scheme has two 'tiers'. The 'Mid Tier' offers five year agreements aimed at delivering environmental gains more widely across a farm with greater opportunities for coordination between clusters or groups of farmers. Although applications are still competitive, this scheme is open to all farmers across England and aims to improve local biodiversity, the historic environment, and water quality. Capital for items to assist with land management options, such as fencing, hedge repair, stone wall building, improving water quality, and planting woodland is also available. The 'Higher Tier' of CS, with agreements typically signed up for ten years, targets the most important wildlife sites including those designated as SSSI. They can include habitat restoration and re-creation – anything from reedbeds to flower-rich hay meadows and woodland; improved habitat management to benefit wildlife; specific habitat provision for key farmland species; and improvements to historic features.

There is concern from some farmers that the CS scheme might be too complex and bureaucratic. A simple example, taking one of its very many options (the guidance runs to hundreds of pages), would be sowing a 'nectar flower mix' to provide areas of flowering plants to boost essential food sources for beneficial insect pollinators including bumblebees, solitary bees, butterflies, and hoverflies. This particular option is available for both Mid Tier and Higher Tier agreements and can be implemented on arable land, on temporary grassland, or in orchards. The rules state that it has to consist of at least four nectar-rich plants and at least two perennial plants (from a stipulated list); that it's sown between certain dates; that half the sown area is cut in spring and the whole area cut in autumn or winter; and that it's not grazed in spring and summer. Some farmers think that the requirements are too rigid; those overseeing the scheme argue that they have to justify taxpayers' money being spent and need rules. CS pays £511 per hectare of land sown with the seed mix. To avoid paying a farmer twice for the same item, if the land is also included by the farmer in his EFA, the Stewardship payment drops to £107 per hectare.

Many measures in Countryside Stewardship come as 'packages' that combine several related options. So, for example, a Mid Tier package for an arable farm on heavy soils growing mainly winter crops on 250 hectares with some high-quality hedgerows between fields would consist of planting up 1.5 hectares with a nectar flower mix, creating one hectare of flower-rich margins around the crop fields, and scattering crop seed (such as barley, triticale, linseed, and sunflower) over five hectares here and there across the farm to encourage seed-eating birds. The annual payment for this package of options would be a very precise £4,505.50.

A new agri-environment climate scheme was launched in Northern Ireland early in 2017. Administered by their Department of Agriculture, Environment and Rural Affairs, it has three 'levels': a Wider Level Scheme aimed at delivering benefits across the wider countryside outside of environmentally designated areas; a Higher Level Scheme primarily aimed at site-specific environmental improvements at important locations and for priority habitats and species; and a Group Level Scheme to support co-operative work by farmers in specific areas, such as river catchments or on common land.

The Scottish Government's scheme doesn't distinguish 'levels' but assesses applications on what environmental value they deliver, covering a broad range of important habitats, species and measures to reduce farming's impact on climate warming, flood reduction and other measures. As in Northern Ireland, most agreements run for five years.

Apart from the very minor advantages for wildlife from farmers implementing the mandatory cross-compliance and greening measures required of them for receiving their BPS payments, it's these agri-environment climate schemes in their differing incarnations that are currently the only effective means of promoting more wildlife-friendly measures on large numbers of farms.

Agri-environment schemes are most certainly not an alternative to farming. Many of the measures in such schemes are not taking land out of food production; they are aimed at modifying farming practices such as applying no fertiliser and pesticides to a cereal crop or re-instating hay meadow management to encourage the return of a wider variety of plants and invertebrates. They might result in lower crop yields but payments are made on the basis of income foregone to reflect this. Recent research suggests that at least some such measures can actually boost crop yields;[xi] an issue that will be returned to in subsequent chapters but which is likely to have implications for some future agri-environment payment rates.

By 2014, just over 3.1 million hectares of land had been entered into agri-environment schemes UK wide.[x] Problems have arisen, though, because of changes made to various schemes over the last decade or so. For instance, with the previous entry level scheme in England closed to applicants (because it delivered little wildlife gain), existing agreement holders will have to apply afresh for a Mid or Higher Tier CS agreement or withdraw completely, thereby jeopardising the small environmental gains that have been achieved on entry level farms and wasting the money invested. Withdrawal is also likely to discourage farmers from applying for a new scheme, especially because they know that restricted budgets mean more competition and less chance of success. Budgetary restrictions for agri-environment schemes – with so much farm support being spent on the BPS subsidy scheme – are a disincentive which needs correcting post CAP if meaningful progress is to be made in returning wildlife to farmland.

Without such schemes, the plight of wildlife on many of the UK's farms, especially lowland farms, would be very much worse than it currently is. In subsequent chapters that deal with creating, recreating, and better looking after wildlife habitats and the species they support, the ways in which these schemes have been used by farmers, and the views of those who have often advised them, will be central. When the UK leaves the EU and each of the four UK countries decides how to allocate its agriculture budget, hard decisions will have to be taken about the level of public support going into farming compared with many other calls on the UK Exchequer.

Will some equivalent of BPS (subsidies) continue to be paid to farmers? If so, should they be restricted to hill and upland farmers, those that farm under the most difficult circumstances? Should a limit be put on the amount that can be claimed per farm? Should the so-called cross-compliance requirements that currently have to be met to receive any subsidy be tightened to expect higher standards of environmental care? Shouldn't the CAP's 'greening' measures be abandoned in favour of much more money being allocated to measures supporting the re-creation and maintenance of existing wildlife habitats on farmland, whether they are called agri-environment schemes (as the CAP jargon dictates) or something else? There is another issue too: should such schemes remain voluntary or be made compulsory, effectively replacing subsidies, thereby ensuring that taxpayers can see tangible benefits for the public money that farmers receive not only in food production but in wildlife too?

It might never have been logical or practical to have one overarching common policy for farming across 28 (soon to be 27) states, but the UK's four countries will shortly be formulating their own agricultural policies. The degree to which these strategies combine environmental protection and provision for wildlife on farmland, both in our uplands and hills and in our lowlands, will be the subject of intense lobbying by Britain's extremely active and voluble conservation organisations whose combined membership is now over six million individuals.

Endnotes

i   europa.eu/pol/agr/index_en.htm.
ii   'Welsh Agricultural Statistics 2014,' Welsh Government.
iii Statement in the Scottish Parliament, 11 June 2014.
iv   'Agriculture: A Partnership between Europe and Farmers,' European Commission, 2014.
v   'Basic Payment Scheme: Rules for 2016,' Rural Payments Agency, 2015.
vi   Communications from Department of Agriculture and Rural Development, Northern Ireland, and Welsh Government Environment and Countryside Department, 2016.
vii Communication from Scottish Government's Department of Rural Affairs, Food and Environment, 2016.
viii Data supplied by Tony Morris, RSPB, based on several published papers.
ix   Defra figures quoted in *Farmers Weekly*, 25 August 2017, page 26.
x   'Green direct payments: implementation choices of nine Member States and their environmental implications,' Kaley Hart. Institute for European Environmental Policy, 2015.
xi   *Farmers Weekly*, 25 August 2017, page 56-57.

CHAPTER 5

BRINGING BACK THE BUZZ

What would be left of our tragedies if an insect were to present us his?

EMILE CIORAN, ROMANIAN PHILOSOPHER AND ESSAYIST

'There's one; do you see it?' asks Henry Edmunds excitedly as he points to a male Adonis Blue flying low over his downland. It's a butterfly that's generally confined to chalk grassland and similar habitats in the south of England and is more vibrantly coloured than I had imagined. Then Henry spies something else flitting from plant to plant in the flower-speckled turf: 'That's a Five Spot Burnet and look, over there near that rockrose, a Brown Argus with a Burnet Companion moth next to it.' I'm rapidly losing track, attempting in vain to scribble down the names as Mr Edmunds, owner of the 1,000 hectare Cholderton Estate near Salisbury in Wiltshire, reels off the name of one butterfly or moth after another. There's so much to see on this lightly grazed, warm, sunny downland slope that I hardly know where to look next as he leads off to spot yet more riches.

'We usually graze here with a few Hampshire Downs sheep but only in winter after all the plants have set seed. We do a bit of scrub control by hand, enough to leave some patches but we don't want it encroaching on too much of the grassland,' says Mr Edmunds. 'As far as we know this downland has never been ploughed so it's been like this perhaps for centuries. And that's how it's going to stay here; you need to have a respect for all life and be a proper steward of your land. Industrial farming has lost that ethos altogether; some farmers would clear all this, spray it with glyphosate to kill the lot, then plough it to grow barley. They're silly asses,' he comments ruefully. Henry Edmunds is not a man to pull his punches.

'This place is just a delight,' he says as we sit surrounded by a cornucopia of rare flowering plants including the blue-flowering spikes of Meadow Clary and butter-yellow Field Fleawort. 'And we have recorded over 450 species of moth and 34 species of butterfly which we think breed on the Estate. Cholderton supports a good population of the rare Brown Hairstreak butterfly in scrub and around the edges of some of our woodland and we have a range of other invertebrates including the rare Hornet Robber Fly,' (a large predatory hunter that eats beetles, grasshoppers, wasps, and other flies including, at times, each other).

Thirty years ago much of the chalk downland at Cholderton was covered in Scots Pine and broadleaved scrub. Since then the trees have been thinned, leaving a few to provide diversity. For the first three or four years of thinning there was little to be seen of any downland flowers. Many areas remained quite bare, or were colonised by fescues and other grasses, but within six years a downland flora was developing which the Estate assisted by grazing only after the flowering plants had set seed. Today this downland in spring has Cowslips and carpets of Milkwort followed by huge areas of Horseshoe Vetch, its yellow flowers buzzing with bees and early season butterflies. In summer a profusion of flowering plants is accompanied by a multitude of flying insects: bumblebees, robber flies, and butterflies such as Marbled Whites, Meadow Browns, and Gatekeepers.

With financial support from the Environmental Stewardship scheme (since superseded by Countryside Stewardship), Henry Edmunds is recreating large areas of downland from fields that had in the past been converted to ryegrass pasture or used for crop growing. It's being restored by seeding it with seed collected from areas of existing downland or by buying in appropriate seed mixes. Some fields that were growing cereals maybe 15 years ago are today supporting up to a hundred species of plants together with diverse populations of invertebrates.

Sustainable farming is the guiding principle of the Cholderton Estate; it's been managed for the past 30 years with the aim of achieving a balance between the demands of modern, highly competitive agriculture and the preservation of the countryside, working with nature rather than against it. The Estate is a registered organic farm and Henry Edmunds' livestock is fed almost entirely on food that is grown on it. It won the RSPB Nature of Farming Award in 2012.

The Estate has 220 hectares of arable crops with oats and barley interspersed on a rotation with Sainfoin-rich grass leys (see Chapter 6), kale, or turnips and an occasional crop of vetches which are extremely vigorous and smother out most 'weeds'. However, like Sainfoin (grown at Cholderton since 1730), they are extremely attractive to a wide range of pollinating insects, bumblebees and honeybees especially. Margins left without sowing crops along the edges of some of the arable fields are allowed to become grassy strips in which arable flowers proliferate and a wide range of insects flourish. The Estate's larger arable fields are divided by 'beetle banks', grassy mounds about two metres wide that run through the middle of the fields and provide a raised, and therefore drier, habitat for overwintering spiders and many insects (including beetles) that will prey on crop

Henry Edmunds amongst his flower and insect-buzzing chalk downland, Cholderton Estate, Wiltshire.

pests such as aphids come spring and summer. These measures are also supported by payments from England's agri-environment scheme.

The problem is that, unlike the farmland on Henry Edmunds' estate, most lowland farms in Britain are not teeming with wild bees, bumblebees, butterflies, moths, and a plethora of other insects. Yet as many such insects are essential for pollinating a great number of farm and horticultural crops, our farmers – and us as consumers – will be in a disastrous fix if we allow them to die out. Pollination is the vital process by which many plants breed, and it's been suggested that up to 84% of EU crops (valued at £12.6 billion) and around 80% of wildflowers depend on pollination by insects.[i]

In Britain the majority of pollination is carried out by bees (wild solitary bees and bumblebees, as well as domesticated honeybees), flies (including hoverflies and bee-flies), butterflies, moths, wasps, and beetles. Commercial crops that benefit from visits by wild insect pollinators include oilseed rape pollinated by short-tongued insects including hoverflies and honeybees as well as bumblebees; field bean flowers most easily accessed by long-tongued species of bumblebee; and strawberry flowers visited by solitary bees, hoverflies, and bumblebees. Tomatoes, aubergines, and peppers require 'buzz pollination' which relies on bumblebees and some solitary bee species producing vibrations using their flight muscles to dislodge the pollen. Apple pollination is carried out predominantly by solitary bees and honeybees. Although our food supply depends on them, agronomists and the companies producing and marketing pesticides (insecticides to kill insects directly or herbicides to kill off the unwanted plants many such insects depend upon) are helping to destroy them. Other factors including climate change, substantial losses of habitat, and invertebrate pathogens (such as Varroa mites that infect bees) are all having an impact.

Characteristically, Henry Edmunds puts the issue rather more bluntly.

'The fertiliser and pesticide companies have too much of a vested interest in intensive farming and too many farmers are spending £250 a hectare spraying their crops. The silly asses are destroying our wildlife and the organic content of farm soils along with it,' he says.

The 2016 State of Nature report concludes that over the long term (1970–2013), of the regularly monitored 517 species of farmland invertebrates, about 20% have declined strongly in number and another 18% have declined moderately while around 27% have increased.[ii] More worryingly, it also found that 50% of the monitored invertebrates declined in recent years (2002–2013) suggesting that the rate at which many invertebrates – groups such as butterflies, moths, bees and dragonflies – declined is actually accelerating.

All of the UK's agri-environment climate schemes include measures to provide or improve habitats for a range of pollinating and other insects, including creating flower-rich crop margins and beetle banks, leaving winter stubble from cereal crops, and growing areas of tall grassy vegetation. Farmers might reasonably be expected to invest in such essential requirements to boost the numbers of pollinating insects for their crops themselves. Instead, and in addition to the BPS direct payments (subsidies) and greening payments they receive, these fundamental requirements are paid for by taxpayers via agri-environment schemes!

But it's not a lot of use 'pepper-potting' bits of good invertebrate habitat on one farm here and there. Such habitat needs joining up so that less mobile invertebrates (and other animals) can move from one good piece of habitat to another. Natural England has started to address this issue by assigning part of its agri-environment scheme to encouraging groups of at least four farmers managing land (normally over 2,000 hectares in total area) to submit landscape-scale projects. To date it's been able to get 1,350 farmers to bring over 270,000 hectares into the scheme.[iii] More such initiatives are needed badly UK-wide.

It's natural to assume that taking land out of crop production to provide insect-attracting crop margins full of flowers and grasses will deplete crop yields on a farm. In fact the opposite is true, even though existing agri-environment schemes currently compensate farmers for such perceived losses. Data from a six year, large-scale study led by Dr Jonathan Storkey of Rothamsted Research on a commercial arable farm with field beans, oilseed rape, and wheat showed that improved pollination because of such margins, combined with natural pest control, led to increased crop yields. Margins took no more than 8% of land at field edges and crop yields increased by up to 35%.[iv]

ATTRACTING THE POLLINATORS

To encourage pollinating insects valuable for farming, the conservation charity Buglife recommends some straightforward measures including:

- Manage existing wildflower-rich grasslands to maximise flowering throughout the season and, if possible, delay cutting hay meadows to allow later flowering species to flower and seed.

- On permanent pastures, implement lighter grazing regimes and/or no grazing periods in summer to increase flowering. Light cattle grazing will generally allow more plants to flower than sheep grazing. Reducing inputs of fertiliser helps too.

- Sowing legume and herbal leys rather than pure grass mixes. Legumes are especially important for bumblebees.

- Establishing flower-rich margins alongside crops to provide additional pollen and nectar resources. Combining permanent wildflower margins with other pollinator-friendly features such as tussocky grass margins, hedgerows and ditches will produce increased benefits for wild pollinators.

- Planting pollen and nectar mixes of plants can provide wonderful and long displays of flowers.

- Taking field corners out of production and leaving them to regenerate with native vegetation, particularly if located alongside hedges, woodlands or ditches. Cutting these areas infrequently (every two or three years) will increase flowering.

- Managing hedges on a two or three year rotation to encourage flowering. Hedge bottom plants provide vital food for pollinators, so flower-rich hedge bottoms need protecting.

- New hedges should use a variety of shrubs to provide a long flowering season.

- Creating or maintaining tall tussocky grass margins to provide important nesting and overwintering areas, cut on rotation or in sections which will provide longer areas of vegetation.

- Keeping earth banks and dry ditches in or around fields. These are often used by small mammals like voles to create their burrows and once abandoned these are often re-used by wild pollinators such as bumblebees to nest. Avoid disturbing such sites between March and October.

- Including some winter stubbles will provide important refuges: the woody plant stems can be used for overwintering.

- Keeping any standing and fallen deadwood and in-field and hedgerow trees to provide nesting areas. Used by overwintering pollinators, decaying wood can provide important breeding sites for hoverflies and other invertebrates.

(Source: 'Helping Pollinators at Farm-scale,' Buglife. https://www. buglife.org.uk/helping-pollinators-at-farm-scale.)

In a distinctly wet part of Britain, Will Dracup has put considerable effort into conserving and improving important wildlife habitats on Broadaford Farm on the edge of Dartmoor. 'It's an 80 hectare family farm my father had before me and we've always been interested in retaining its wildlife as much as we are in farming here. About 49 hectares of it is decent grassland for the 400 or so sheep we have. And we raise suckler cows for about a year on the farm before we sell them off to be fattened on lower ground. We've had an Entry Level Environmental Stewardship agreement with Natural England for several years on the whole farm and now a Higher Level Stewardship agreement on our rhos pasture which mainly follows the shallow valley bottom around the stream here. We've got about 15 hectares of rhos pasture.'

Dartmoor has 20% of England's rhos pasture (boggy ground with tussocks of Purple Moor Grass and rushes) – around 1,200 hectares – which is now becoming a scarce habitat because much of it has been drained, ploughed, and converted into pasture or cropland. Walking over some of this very wet, boggy ground with Mr Dracup we come across lots of the leaves of Devil's-bit Scabious, not yet in flower, and some Meadow Thistle, characterised by its lack of spiny leaves but which will produce large, red-purple flower heads in midsummer. Both plants, and a few pale pink Heath Spotted Orchids are hemmed in by tussocks of tough, pale coloured Purple Moor Grass (it gets its name from its purple flower spikes) which are not easy to walk through. There are much larger tussocks too; these are of the aptly named Greater Tussock Sedge, large clumps well over a metre high of an ultra toughie whose spread can be controlled only by burning it now and then.

But Will Dracup's rhos pastures are not just interesting for the wetland plants they nurture; they are home to a wide range of insects, some of them distinctly uncommon. While attractive black and white Marbled White butterflies can

Careful management. Rhos pasture in Devon where insects including
Marsh Fritillary butterflies thrive on Broadaford Farm, Dartmoor.

be abundant here, it's the much rarer Marsh Fritillary, a chestnut and black-chequered butterfly which has declined by nearly 50% since the 1970s that's one of Mr Dracup's specialities.

'The caterpillar's main food plant is Devil's bit Scabious and Will Dracup's rhos pastures are ideal for it,' comments Jenny Plackett, Butterfly Conservation's Senior Regional Officer for the South West. 'Over winter these caterpillars seek the shelter of dense grass tussocks in which they weave a small web to protect themselves so the dense moor grass here fits its needs perfectly. Another rarity, the Narrow-bordered Bee Hawkmoth – a bumblebee lookalike and mimic – whose larvae also feed on Devil's bit Scabious, occurs here too, often in association with the fritillaries. It's a hawkmoth that has long been declining and now has a very scattered distribution in parts of the UK. This is an important site for it.'

'I've fenced off the rhos pastures from the sheep-grazed fields so I can control the grazing on them. The scheme helped me pay for that,' says Mr Dracup as we stand in the middle of a particularly wet part where even the moor grass is struggling a little. 'In winter there's no grazing on them at all, then I put cattle in sometime in early June usually and they'll graze the moor grass and stay there until autumn. But the timing depends on the season; I look at the vegetation to see if it's right. And I've taken advice from Butterfly Conservation; they've helped a lot. Sheep won't graze in here, it's too tough and wet for them but the cattle tramp the vegetation so it keeps it a bit open. We've done some scrub clearance, willow mostly. It can spread quite fast here so we sometimes have to get some of that cleared and burn part of the tussock sedge as well.'

'Will Dracup has a personal interest in his farm's wildlife,' comments Jenny Plackett. 'He promotes it with schoolchildren from city schools he encourages to visit the farm. Broadaford Farm also has Small Pearl Bordered Fritillary in the rhos pastures; its caterpillar feeds on Marsh Violets but it also occurs on some bracken-covered moorland slopes around Dartmoor where it depends on Common Dog Violet instead.'

'There's a pride with many farmers in the species they still have on their farms in this part of South West England. I work with them and give them advice on how to manage their land to retain or increase what they have and tell them about the financial support they can get through the Countryside Stewardship scheme to help,' says Ms Plackett.

But it's also the approach that an adviser takes when they're talking to a farmer that makes all the difference. 'We get some here,' says Will Dracup. 'A bit like little Hitlers that come and tell you what you must do and what you shouldn't do. But Jenny's not like that. She works with us; she understands how we farm, what the pressures are that we're under too. That's how you get people on side.'

As well as the bovine TB/badgers issue (Chapter 11), another farming issue has stirred substantial public controversy over the last few years. It's the concern about the possible relationship between the decline of bees and the use of insecticide sprays, particularly neonicotinoids (NNIs). Their use has fuelled public anxiety and anger because of the way many farmers – and the insecticide manufacturers and marketers – appear to be determined to argue that they are essential and are not damaging in the face of mounting worries about the continuing decline of bees and other insects.

In the UK, five NNIs are authorised for use in agriculture; they are used as seed treatments for cereals, sugar beet, and oilseed rape so that the seed itself and early growth of the developing plant is protected against insect pests. They are also used as foliar sprays on apples, pears, and a range of glasshouse crops. Seed NNI treatments on arable crops have become increasingly unsettling to beekeepers and bee researchers in recent years, with many experts suspecting that their use is connected to current bee declines. Oilseed rape, the commonly grown crop for which farmers find NNI seed dressings particularly advantageous for killing the very damaging cabbage stem leaf beetle, attracts a number of insects including solitary bees, bumblebees, hoverflies, and honeybees.

In 2013, the European Commission (EC) banned the use of the three most commonly used NNIs as seed coatings, initially for two years but subsequently extended. Having undertaken a review of the latest scientific evidence of harm to pollinators, on 23 March 2017 the EC proposed draft regulations to ban neonicotinoids permanently. An EU-wide decision is expected in 2018. The EC position is that safe use must be demonstrated, not unsafe use, a view that's surely hard to argue with.

In the meantime, emergency authorisations to use these insecticides can be allowed and have been granted in restricted areas of eastern England.

NNIs aren't essential to combat this damaging crop pest; several factors can impact on crop yields and can vary season to season. In 2015 an estimated 17% of growers suffered oilseed rape losses due to the beetle, with an estimated 16,000 hectares lost. That is just 3% of the total crop area, in spite of the NNI ban which farmers railed about.[v] However, the study also found that very much larger than usual insecticide (mainly pyrethroid) doses had been used on the crops to protect against actual or 'predicted' attacks by the pest.

'Based on the body of evidence, we can see that it is absolutely correct to take a precautionary approach and ban these chemicals,' the European Environment Agency's Executive Director, Professor Jacqueline McGlade, said in 2013. 'France has banned some of these chemicals on sunflower and maize since 2004, and it seems productivity has not been affected: 2007 was France's best year for the yield of these crops for over a decade. Also, any economic analysis should consider the almost immeasurable value of pollination carried out by honeybees and other wild bees. Indeed, continuing to use these chemicals would risk a vital service that underpins European agriculture.'

Neonicotinoids act by affecting the central nervous system of insects, leading to their eventual paralysis and death. This specific neural pathway is more abundant in insects than warm-blooded animals, so these insecticides are selectively more toxic to insects than mammals. Bees happen to have more receptors than other insects, as well as more learning and memory genes, but possess fewer genes for detoxification compared to other insects, making them more susceptible. While the older organophosphate and carbamate insecticides tend to degrade quite rapidly in the environment, NNIs are more persistent. Some can persist for months or years in soil and may leach into groundwater under some conditions. Biologically active at very low concentrations, NNIs can be applied at much lower volumes in the field than the older groups of insecticides; doses of a few grams, rather than kilos, per hectare.

Originally welcomed as much safer for humans, livestock, and birds than other insecticides, seed treatments were seen as a more effective method of targeting pests than spraying crop foliage, and more environmentally-friendly. But they pose other risks because they persist in crops and soil, they are toxic at tiny concentrations, and they can be present in nectar and pollen (translocated from the leaves), the very parts designed by plants to attract pollinating insects, bees included.

The salient question is: are NNIs responsible for killing pollinating insects, bees in particular, many species of which are in decline? The issue is complex because many pollinating insects are declining for a number of different reasons: habitat loss, particularly on farmland; disease; insecticide use; adverse weather; and perhaps other factors as well, all or some of which might be interacting in ways we don't understand. For instance, both pesticide exposure and food shortages can impair immune responses, rendering bees more susceptible to parasite infections while exposure to some fungicides can greatly increase the toxicity of insecticides.

ANATOMY OF A KILLER

Work on the development of neonicotinoids (chemically related to nicotine) began in the 1980s, and they were introduced onto the market in the early 1990s. Compared to organophosphate and carbamate insecticides, neonicotinoids cause less toxicity in birds and mammals than insects. One type of NNI, imidacloprid, is the most widely used insecticide in the world and, overall, NNIs comprise nearly a third, by value, of the international pesticide market.[iv] They are used to control a variety of pests, especially sap-feeding insects such as aphids on cereals, and root-feeding grubs. They are systemic pesticides, meaning that unlike contact pesticides, which remain on the surface of the treated foliage, they are taken up by the plant and transported to all the tissues (leaves, flowers, roots, and stems, as well as pollen and nectar).

Products containing NNIs are commonly applied as seed coatings or sprayed onto crop foliage and the insecticide remains active in the plant for many weeks, protecting the crop season-long. Some garden insecticides contain them too. However, since they get into pollen and nectar, pollinating insects ingest them. These and other invertebrates, including aquatic species, might also get exposed to them from plant exudates, from dust created by seed planting machines, or from contaminated soil and water the insecticide easily leaches into. Using them as seed coatings is akin to people taking antibiotics as an insurance policy to try to avoid getting ill. Critics say that this is the kind of approach that eventually will build up resistance in the very pests the chemicals are designed to kill.

No one doubts that many bees and other pollinators have been in decline for decades in the UK, yet there's no agreement on whether NNIs are at least partly, or even largely responsible; some studies have concluded that they harm bees, others have found little effect. A major review published in 2014 found that there was limited evidence of an effect but that there were substantial gaps in information about how these pesticides might affect bees.[vi] A more recent study found that oilseed rape seeds coated with a combination of the neonicotinoid, clothianidin, and the non-systemic synthetic pyrethroid, β-cyfluthrin, reduced wild bee density, solitary bee nesting frequency, and bumblebee colony growth and reproduction under field conditions.[vii] It concluded that such insecticidal use can pose a substantial risk to wild bees on farmland, and that the contribution of pesticides to the global decline of wild bees may have been underestimated.

Buff-tailed Bumblebee (courtesy of Tony Wills).

The lack of a significant response in honeybee colonies suggested that reported pesticide effects on wild bees cannot always be extrapolated to honeybees.

A very recent study related 18 years of UK national wild bee distribution data for 62 different species to the amounts of NNI used in oilseed rape crops. It found that sub-lethal impacts of NNI exposure can be linked to large-scale population declines of wild bee species in England, with these effects being strongest for species that are known to forage on oilseed rape crops.[viii] It acknowledged that short-term laboratory studies on commercially bred species (principally honeybees and bumblebees) have identified sub-lethal effects but, until recently, that there has been no strong evidence linking these insecticides to losses of the majority of wild bee species.

While the average decline in population across all 62 species was 7%, the average decline among the 34 species that forage on oilseed rape was higher at 10%, and some fared much worse. Five of the 62 species – including the Spined Mason Bee and the Furrow Bee – declined by 20% or more, and the worst affected bees declined by 30%. 'Prior to our study,' says Dr Ben Woodcock, of the Centre for Ecology and Hydrology, who co-led the research, 'people had an idea that something might be happening, but no-one had an idea of the scale. Our results show that it's long-term, it's large scale, and it's many more species than we knew about before.

'To put the research presented in our paper into context it needs to be considered as complimentary to previous work. Previous research manipulating doses of NNIs has been crucial in identifying their impacts on bees but these studies are limited because they are short term, all done in less than a year' – bees have one generation per year – 'they are small scale, often in a lab or a restricted land area; and they are limited to three main species – honeybees, Buff-tailed Bumblebee and the Red Mason Bee – simply because these can more or less be artificially bred and can be used in an experimental environment.

'Our work compliments these studies that have identified the mechanism of impact on bees and extends it to whole faunas. Although our study is correlative we look at longer time periods' – 18 years – 'we look at a national scale' – over 400 one kilometre grid square across England – 'we look at 62 species. So we have shown that previously identified effects of NNIs on bees scale up to national populations of many species over extended periods of time.

'In my opinion, NNIs are a problem and do have negative effects on bees. Unfortunately pesticides are designed to kill and the question is what is the alternative; what will the alternative do to not just bees, but everything else; and what are the financial impacts versus the impacts on wildlife? The bee declines are subtle in some species; worse in others and I suspect that impacts on wild bees come from several sources: habitat loss and fragmentation, disease, climate change and other insecticides that may be interacting with NNI exposure to affect their populations. Healthy bee populations in places with lots of alternative foraging resources and nesting sites are probably fairly robust, while those in areas intensively-farmed and devoid of more natural habitat may easily be tipped over the edge by exposure to NNIs.'

The cause of honeybee colony collapse disorder (CCD) isn't clear but NNIs have sometimes been implicated and might play a role. CCD results in the sudden death of a colony with a lack of healthy adult bees inside the hive. It appears to affect the adult bees' ability to navigate; they leave the hive to find pollen and never return. Honey and pollen are usually present in the hive, and there is often evidence of recent brood rearing. In some cases the queen and a small number of survivor bees may remain in the brood nest. CCD is also characterised by delayed robbing of the honey in the dead colonies by other, healthy bee colonies in the immediate area, as well as slower than normal invasion by common pests such as wax moths and small hive beetles. But it's a phenomenon that seems to have been recorded occasionally for centuries, and other causes including parasites and predators, poor nutrition in inclement weather, and even genetic factors could all play a role.

There is recent evidence that NNIs can reduce a bee's ability to carry out what's known as 'buzz pollination' for plants such as tomatoes, potatoes, and aubergines that need vibration to shake pollen out of their anthers, a bit like shaking salt out of a salt cellar.[ix] Bumblebees gradually learn the technique of buzz pollinating, thereby improving their pollen collection rates as they get more experienced. Dr Penelope Whitehorn's recent research at the University of Stirling found that bumblebees fed field-used levels of thiamethoxam, a common NNI, failed to collect more pollen over time compared with bees not fed the chemical. It seems that the NNI-affected bumblebees had their learning impeded, which is bad news for the pollination of certain crops. Added to recent research showing that honeybees' ability to recall where they are and to navigate back to their nests are both reduced when exposed to the pesticides, it suggests that NNIs result in poorer foraging and pollination.

Almost all studies have concentrated on NNIs in crops and their impact on certain insects, bees especially. Consequently, there has been a lack of information about any impact that such residues in wild plant pollen and nectar might have on bees and other pollinators where flower-rich margins are encouraged (usually funded by agri-environment climate schemes) adjacent to NNI treated crops. One such study found that in summer, NNIs in the pollen and nectar of wildflowers growing in arable field margins can be at concentrations even higher than those found in the adjacent treated crop.[x] Indeed, the large majority (97%) of NNIs brought back in pollen to honey bee hives in arable landscapes was from wildflowers, not from the crops themselves. Other field studies have been based on the premise that exposure to NNIs would occur only during the flowering period of farm crops and that such exposure might be diluted by bees also foraging on untreated wildflowers. That seems now to be incorrect; their exposure is likely to be higher and more prolonged than currently recognised due to widespread contamination of wild plants growing near treated farm crops.

There are also studies correlating the area of farmland where NNIs are used with the decline of 15 of our 17 species of farmland butterflies.[xi] These declines have largely occurred in England where NNI usage is at its highest; in Scotland, where their use is comparatively low, butterfly numbers are stable. Studies like these don't necessarily prove cause and effect; some other as yet unidentified factors could also be involved. But is it merely a coincidence?

Professor Ian Boyd, Chief Scientific Adviser at Defra, remains skeptical that NNIs are to blame for bee declines: 'Very few studies have been specifically designed to answer the question as to whether these insecticides affect the populations of non-target insect species when used in normal circumstances. This is because these studies are very difficult to carry out. Those few studies that have been conducted to answer this question generally have low statistical capacity to demonstrate effects. Sometimes they show there are effects and sometimes they don't. The effects appear to be small (and therefore difficult to detect) compared with those associated with weather variation, landscape management, and the effects of pests and disease on insects like honey bees. Neonicotinoids are part of a wider system of farming that probably has a much greater impact overall. Focus needs to be on that "system" and not on one component of it. And it's also the case that the consequences of outright banning of technologies can lead to adverse outcomes. In the case of the recent suspension of neonicotinoids, we know that this has led to the increased use of older, less effective but often more environmentally damaging chemicals. Managing these adverse outcomes is often easier said than done.'

A BUZZING LIFESTYLE

Bees are closely related to wasps and ants. They are known for their role in pollination and, in the case of the best-known bee species, the European Honeybee, for producing honey and beeswax. There are nearly 20,000 known species of bees in the world (and several hundred in the UK) including bumblebees, carpenter bees, mason bees, digger bees, sweat bees, and leafcutter bees. They are found on every continent except Antarctica and in every habitat that contains insect-pollinated flowering plants.

Colonial species live in colonies that contain the queen bee, thousands of worker bees, and a few hundred drones. Workers and the queen bee are both female, but only the queen bee can reproduce. All drones are male. Worker bees clean the hive, collect pollen and nectar to feed the colony, and take care of the offspring. Drones have a far more pleasant life; their job is to mate with the queen, while the queen's only job is to lay eggs. Honeybees are particularly sociable and can live in groups of up to 50,000 in a single hive while bumblebees (24 British species) live in smaller groups of 50 to 150. Both types depend on queen bees who lay all of the eggs for the colony. In contrast, solitary bees often make burrows underground or live on their own in quarry faces, old wood, and even masonry.

Bees are adapted for feeding on nectar and pollen, the former primarily as an energy source and the latter primarily for protein and other nutrients. Most pollen is used as food for their larvae. Bee pollination is important both ecologically and commercially; the decline in wild bees has increased the value of pollination by commercially-managed hives of honeybees. Pollinators such as bees, birds, and bats are essential for 35% of the world's crop production, guaranteeing the output of 87 of the leading food crops worldwide.[xii]

On the other hand, several experts place NNIs firmly in the dock. 'Doing full-scale field trials with NNIs and bees is very time consuming and expensive because control areas free from pesticides have to be massive to ensure the bees don't stray out of them. However, the evidence for bumblebees is very clear: exposure to field-realistic levels and in natural scenarios harms bumblebee colonies. The evidence for honeybees is less clear: NNIs certainly impair their learning, navigation, egg-laying, and immune system but there has been no convincing large-scale field trial showing harm,' says Professor Dave Goulson of the University of Sussex.

Professor Goulson also disputes the concern that farmers who are denied these insecticides will turn to using others that might be more wildlife-dangerous. 'Alternative, more harmful chemicals are not available for farmers. They resort to pyrethroids which break down much more quickly in the environment than NNIs. It's a shame, though, that farmers don't explore alternatives to pesticides such as trap crops, encouraging biocontrol agents, and others,' he comments. One obvious way of reducing insect exposure to any of these pesticides is to encourage far more farmers to convert to organic farming (see Chapter 10). Organic farmers can use naturally occurring pesticides if they need to, some of which are very toxic, but not synthesised pesticides such as NNIs and others.

Insecticides sprayed on to plant foliage (not systemic insecticides such as NNIs coating oilseed rape seed) have less impact on wild bees. Existing agricultural codes of practice (which are not mandatory) minimise the risk of exposure for domesticated and wild bees by limiting the application times of these insecticides to periods of low bee activity, particularly the evening or early morning. Such codes of practice were developed principally to protect honeybees that forage over a well-defined daily feeding period. However, other wild bee species, in particular bumblebees, may forage over a larger proportion of the day and so may be more likely to suffer mortality from foliar insecticides even where codes of practice to protect them are adhered to.

The precautionary approach taken by the European Commission to ban the use of NNIs seems to be common sense. When the UK leaves the EU the decision on whether to use them will be taken by the four governments of the UK. However, the bulk of their use is in England where the farming industry, the pesticide manufacturers, and the NFU will undoubtedly lobby hard for their continued use. If the objective evidence indicates a link between their use and pollinating insect declines, will the UK Government also adopt precautions, or throw caution to the farming wind? In any case, if a permanent NNI ban is imposed in all EU countries, British farmers post Brexit are unlikely to be able to sell any NNI-treated produce Europe-wide.

It isn't only bees that are in trouble on lowland farmland. Butterflies and moths are not doing well either. 'We're not sure what the cause is but a whole set of our more common farmland butterflies are declining,' says Martin Warren, former Chief Executive of the charity Butterfly Conservation. 'Species such as Gatekeeper, Meadow Brown, Orange Tip and the skippers, all of them generalists in terms of the plants they need to feed on, are suffering. Some of these overwinter on dead plant flower-heads so field edge vegetation needs to be left uncut, it needs a bit more untidiness I suppose around the base of hedges for example. These sorts of places are often cut too frequently on many farms. Hedges, too, shouldn't be cut so often and, when they are cut, not all of those on a farm every year.

'We've also seen a decline in the more common species of moths, especially in South East England which is where the most intensive farming is, though the picture might be complicated by the effect of climate warming. But we desperately need to see natural vegetation corridors connecting up good habitat from one farm to another so mobile species can actually move from place to place. That's essential. It's why targeting of agri-environment schemes is so important,' he adds.

Hope Farm in Cambridgeshire is managed by the RSPB as a commercial enterprise, with numerous measures to cater for wildlife. Here, butterflies are systematically monitored weekly from April until the end of September to provide a yearly total for each species. Annual comparisons can then be made with national trends. Using the same suite of 21 species, the Hope Farm index shows that overall, butterfly numbers more than doubled (2.5 on the left scale) in abundance since the baseline year (1 on the left scale) of 2001, although with some annual fluctuation because of poor weather. The national trend for England has been a slight decline over the same period (and a substantial decline over the previous decade).

Hope Farm has developed flower-rich margins around crops which include plants that produce plenty of pollen and nectar; wild bird cover crops; and Skylark plots (patches of unplanted ground in the middle of crops), together with reduced hedgerow management. All of these measures have helped boost butterfly numbers and plenty of other invertebrates too, showing what can be achieved on a profitable, lowland Cambridgeshire farm.

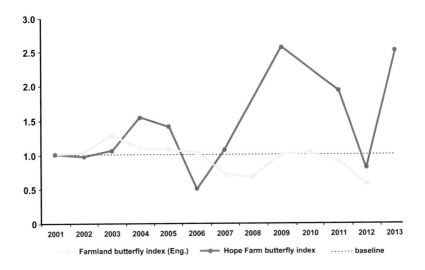

Hope For Butterflies (source RSPB)

Studies have found that crop margins established by sowing with a grass and wildflower seed mix attracted more adult butterflies, while larger and more species-rich invertebrate populations survived better on uncut margins than on those that were mowed annually, though in the long term such gains were likely to be reversed as scrub encroached.[xiii] Consequently, rotational mowing, leaving some areas of longer vegetation (on the outer edge of the field most distant from the crop maybe) and cutting other areas after flowering, is probably ideal. Even better is to retain a hedge adjacent to the arable margin; that benefits other invertebrates including pseudoscorpions – tiny scorpion lookalikes that prey on smaller invertebrates – as well as a range of farmland birds that use the hedge for breeding and the flower-rich margin for feeding.

Farming is needed to maintain habitats for many butterflies in the UK, yet agricultural intensification is a principal cause of their long-term decline. Ironic, certainly, but it's a clear conclusion from a detailed analysis of the state of Britain's butterflies, which showed that in England there was a 57% decline in butterflies on farmland between 1990 and 2014, continuing a trend that began much earlier and is more noticeable in the south of Britain than in the north and west.[xiv] Much the same has happened to moths, suggesting that both butterflies and moths are suffering the most where farmland is managed intensively and leaves little room for wildlife habitats to prosper. Field corners, banks, wide hedges, and many other bits of land on a farm not managed for growing crops are often places which support common wayside flowers such as Cow Parsley, violets, White Dead-Nettle, and dandelions, that can be crucial for newly emerged queen bumblebees, solitary bees, and hoverflies. Odd bits of unused ground with some tall weeds – Hogweed, teasels, thistles, and willow herbs – can often be some of the best habitats for pollinators around the farm.

Nor is it only butterflies of lowland habitats that are suffering. The Northern Brown Argus, a brown butterfly spotted with orange is also at risk. It is an inhabitant of well drained, lightly grazed or ungrazed grasslands, often on limestone, in eastern Scotland and northern England, much of it on farmland. It has declined in abundance and has been lost from some of its historic locations over the last few decades. Light grazing by livestock of the grasslands it inhabits is ideal but heavier grazing depletes its numbers; cutting down scattered patches of scrub (which it needs for shelter) on outbye land may also have contributed to its decline.

Britain's only true high mountain butterfly, the Mountain Ringlet, another predominantly brown species, is also in decline in its habitat of the high Scottish mountains and the Lake District hills, though it's not clear why. Its caterpillars feed on Mat Grass, a very common plant across the British uplands, while the adults have a liking for hawkweeds, Thyme, and Tormentil, also very common upland plants. Conversely, the Scotch Argus, the only other northern species, has increased substantially in numbers in the tall, often boggy grassland habitat it prefers, where its food plants are abundant. Teasing out the reasons for these very different changes in fortunes isn't at all easy. Subtle alterations in how farmland habitats are managed, small changes in the density of livestock grazing mountain land, the effect of climate warming, seasonal weather variations, and other factors might all be at play and affect different species in very different ways.

Implementing relevant agri-environment measures on farms certainly seems to bolster their invertebrate populations. A three year study at six locations in southern and eastern England compared the efficacy of different Environmental

Stewardship options for field margins on arable land in enhancing the abundance and diversity of flowering plants and foraging bumble bees. It was discovered that uncropped field margins sown with mixtures containing nectar and pollen-producing plants were more effective in providing bumblebee forage than margins sown with a grass mixture alone.[xv] A planted mixture of agricultural legumes such as Red Clover did even better. It established quickly and attracted on average the highest total abundance and diversity of bumblebees including the rare Large Garden Bumblebee and the Large Carder Bee. Any seed mixtures, though, needed to contain a wide range of species (preferably including thistles, plants not often welcomed by farmers) in order to provide bee foraging throughout the season and not just for a short time in summer.

A study of 32 different locations across England with a mix of field margins found that bumblebee numbers in July and August were significantly higher on margins planted with a range of pollen and nectar-producing plants.[xvi] They averaged 86 per 100 metres of margin compared with 43 in natural wildflower margins, just six per 100 metres in mature grass-only margins, and eight in recently-sown grass margins. Bees were virtually absent from the cereal crop itself (averaging only 0.2 per 100 metres). The number of different bumblebee species was much higher on margins sown with either wildflowers or the pollen and nectar mix than on any other margins.

The implication of these studies is that agri-environment schemes containing options for creating flower-rich margins around arable crops (whether sown by seed or allowed to grow up naturally) are of considerable benefit to bumblebees and are probably valuable for many other invertebrates too.

NNIs might well be largely responsible for the serious decline in recent years of many bee species – maybe other invertebrates too – but they are just one part of an intensive lowland farming system that is elbowing out more and more wildlife. Ensuring that the farming industry has the means to counter damaging crop diseases and pest infestations is essential (whether by mechanical means as adopted by most organic farmers or using synthetic chemicals), yet the means of doing so must have a minimal impact on other invertebrates, many of which are essential for crop – and wild plant – pollination. It is something that pharmaceutical companies, and those farm advisers that many farmers rely on for advice on the use of pesticides, should consider as their mantra. It certainly does not appear to be so at present!

Agri-environment schemes are helping to re-establish viable invertebrate populations locally on those lowland farms taking up some of their available measures. Retaining hedges and the few traditional, flower-rich hay meadows and pastures that still exist; sensitively managing areas of downland and heath; tolerating odd bits of overgrown and unfarmed land; leaving patches of scrub and woodland; sowing wildflower and grass margins alongside crops; creating beetle banks; retaining crop stubbles overwinter; and much more all benefit a plethora of invertebrates, plants, and animals. Many farmers are implementing these measures; more wish to do so but are discouraged because agri-environment budgets don't stretch far enough and entry is competitive.

In our uplands, moors, and mountains, livestock grazing numbers are often too high to allow any vegetation much above billiard table height to prosper, thereby limiting its attractiveness to many invertebrates. Some of the steepest, often rocky and inherently dangerous mountain slopes are, though, no longer being grazed. This is encouraging a better mix of more rank vegetation to develop; areas of scrubby broadleaved woodland, scattered trees, some bracken, heathland, and rank grassland, will aid many invertebrates and much other wildlife into the bargain. Agri-environment schemes are promoting this change but much of it is due to the movement, well over a decade back, from livestock headage based subsidies to area subsidies reducing sheep numbers overall and encouraging farmers to re-assess whether it's worth gathering sheep (for shearing, lambing and other purposes) on some of this terrain.

However, it's essential that habitats of value to invertebrates – whether newly created or existing – are connected farm to farm, especially in the lowlands but in upland areas too, in order to enable mobile species to move in response to changing climatic conditions. That can only be achieved by considerably increasing the take-up by farmers, especially by groups of adjacent farmers, of agri-environment measures – or whatever similar schemes replace them post CAP – widely across all of our farmed countryside.

Endnotes

i Pollinators and farming. Buglife.

ii 'State of Nature 2016,' D. B. Hayhow et al. The State of Nature Partnership, 2016.

iii *Farmers Weekly*, 1 September 2017, page 7.

iv 'Making room for pollinators need not harm farm productivity,' Bees Needs 2017- Campaign for the Farmed Environment, July 2017.

v 'An interim impact assessment of the neonicotinoid seed treatment ban on oilseed rape production in England,' Charles Scott and Paul Bilsborrow. Newcastle University Rural Business research, August 2015.

vi 'A restatement of the natural science evidence base concerning neonicotinoid insecticides and insect pollinators,' Charles Godfray et al, 2014. Proceedings of the Royal Society B, 281: 20140558.

vii 'Seed coating with a neonicotinoid insecticide negatively affects wild bees,' Maj Rundlof et al, 2015 . *Nature* 521: 77-80.

viii 'Impacts of neonicotinoid use on long-term population changes in wild bees in England,' Ben Wood cock et al, 2016. *Nature Communications*, 7; article no. 12459.

ix Preliminary research results by Dr Penelope Whitehorn, University of Stirling, 2016/7.

x 'Neonicotinoid residues in wild flowers, a potential route of chronic exposure for bees,' Cristina Botías et al, 2015. *Environmental Science and Technology*, DOI: 10.1021/acs.est.5b03459.

xi 'Are neonicotinoid insecticides driving declines of widespread butterflies?' Andre S. Gilburn et al, 2015 . *Peer J*, DOI: IO.7717peerj.1402.

xii 'Pollinators help one-third of world's crop production, says new study,' Sarah Yang, 2006.

xiii 'How can field margin management contribute to invertebrate biodiversity?' by Ruth Feber et al. *Wild life Conservation on Farmland, Volume 1*. Eds. Macdonald & Feber. Oxford University Press, 2015.

xiv 'The State of the UK's Butterflies 2015,' Butterfly Conservation, 2016.

xv ' Comparing the efficacy of agri-environment schemes to enhance bumble bee abundance and diversity on arable field margins,' C. Carvell et al, 2007. *Journal of Applied Ecology*, 44: 29.

xvi 'Effectiveness of new agri-environment schemes in providing foraging resources for bumblebees in intensively farmed landscapes,' Pywell et al, 2006. *Biological Conservation*, 129: 192.

CHAPTER 6

WHAT ABOUT THE WEEDS?

When weeds go to heaven, I suppose they will be flowers.

LUCY MAUD MONTGOMERY (1874–1942), *THE STORY GIRL*

Weeds already are flowers; they don't need to be in heaven to become so! The term 'weed' is a human construct: a plant considered undesirable in a particular situation, 'a plant in the wrong place'. Gardeners and farmers abhor them. Historically, they had to be kept in check or eliminated in a farm crop, as in gardens, by pulling them out or by regular hoeing of the soil to chop them off.

Today, on most farms, selective herbicide sprays do the job much more efficiently. Flowers – scarlet poppies, blue Cornflowers, vibrant yellow Corn Marigolds, and many others – that for millennia dotted farm crops with splatters of colour are today rarely seen. In fact, these arable flowers are some of our most endangered plants. The UK's Biodiversity Action Plan – overseen by the Government, and which sets an agenda for conservation in the 21st century – identifies arable field margins among the UK's highest-priority habitats. Of the 62 flowering plants selected for most urgent action in the Plan, 12 are plants from arable fields.

But they are not about to die out everywhere. Jake Freestone, Farms Manager on the Overbury Estate in the Cotswolds, is nurturing a huge area of these arable flowers, putting the colour back into what would otherwise be a uniform landscape of greens and yellows as his spring barley, oilseed rape, peas, and beans mature through summer. Mr Freestone has created wide, flower-rich margins along the edges of many of his crops, encouraging dormant flowers to grow and bloom.

I'm with John Clarke and Jenny Phelps, walking along a farm track at the edge of such a colourful margin. John is a local ecologist who has studied these margins

*A riot of poppies and many more flowers besides. Arable field margin
on the Overbury Estate, Gloucestershire (courtesy of John Clarke).*

and other wildlife-rich areas on the Estate, and Jenny is Gloucestershire Farming & Wildlife Advisory Group's Senior Farm Conservation Advisor. The margin is a delight. Common Poppy, its papery scarlet petals flexing in the breeze, is by far the most obvious flower. However, John tells me that there are three other poppy species here that are not so obvious, including Rough Poppy, as well as other arable flowers such as Cornflowers and pinkish-white Night-flowering Catchfly which, as its name suggests, flowers at night. It's a strategy to attract nocturnal moths to pollinate it! The strangely named Venus's Looking Glass is here too; it is named after the shining oval fruits located inside the seed-capsule which are said to resemble brilliantly polished brass hand-mirrors.

'Once the flowers in the arable margins have set seed in the autumn we top them' – mow them lightly – 'so that pernicious weeds like Charlock, Wild Oats and Black Grass don't start getting the upper hand and move into the adjacent crop where we don't want them,' says Jake Freestone. 'Sometimes we have to cut them earlier in the season but the wild flowers are very resilient. We spray the margins in the autumn to kill everything off and either plough them when we plough the adjacent crop or disc harrow them to just disturb the soil surface. Then the wild plants re-grow again naturally in spring. They always seem to do well. After our existing Stewardship agreement ends in 2020 we hope to enter the new Countryside Stewardship scheme but much depends on the payment levels' – to cover the income foregone for not cropping the margins – 'and how easy it'll be to get in. It's a points system and it's pretty competitive. So we can't be sure.'

THE DEL BOY FLOWERS

A century on from the death-ridden French and Belgian battlefields of World War I, poppies represent the terrible losses of that appalling conflict and the hope for new life beyond the carnage. Their use was inspired by the poem 'In Flanders Fields', written in 1915 by Canadian physician Lieutenant-Colonel John McCrae. It refers to the many poppies that were the first flowers to grow in the churned-up earth of soldiers' graves in Flanders. Poppies and other arable flowers like them briefly flourish where ground is turned over before disappearing from sight again for another year. Their seeds wait for disturbance before these scarlet beauties will germinate, grow rapidly, then flower and produce more seed to fall on the soil below. So on farmland the soil must be disturbed, by shallow harrowing for instance, to encourage them to grow.

The Common Poppy, the Cornflower, Corn Marigold, the now almost-extinct Corncockle, and many others, have their origins in the eastern Mediterranean. These plants first found their ideal niche in the earliest farmed fields before 6,000 BC as farming spread across Europe. The seeds hitched a ride, were harvested together with barley and wheat grains, and then accidentally sown wherever the migrant farmers stopped, eventually making their way into Britain across the marshy land that then connected us to the European continent.

It's a different take on the poppy but, like all of these arable field flowers, they are what Richard Moyse, Manager of Plantlife's Ranscombe Farm Reserve in Kent, calls 'the Del-Boy Trotters' of the plant world. 'They are the duckers-and-divers, briefly flourishing where ground is turned over before dropping from sight again. Many of these plants, poppies included, make their best living in arable fields where the ground is in regular and repeated motion like Del-Boy in the continual hustle-and-bustle of a South London market. The poppy is a wide-boy, a beloved British stereotype whose lineage includes Falstaff, the Artful Dodger, and Dad's Army's Private Walker.'

The soil in arable, flower-rich margins like these has to be harrowed or ploughed each spring because most of the flowering plants require soil disturbance to encourage them to germinate. The margins are then left unsprayed with the pesticides used on the adjacent crops through summer, and are never fertilised. Allowing the mix of flowers and grasses to mature and set seed is vital for flowering the following year, as most of them are annual plants that can't compete with today's dense, fertilised, and pesticide-sprayed crops.

Far from marginal. Paul Simpson at Newlands Farm, Dorset in one of his unsprayed wild flower margins around a barley crop.

These margins are not just rich in plants of course. John Clarke has recorded an array of butterfly species here including Common Blue, Marbled White, and some of the skippers, plus a plethora of bees and bumblebees. There are masses of other invertebrates too: grasshoppers, beetles, hoverflies, and more. Seed-eating birds such as Linnets, Goldfinches, Yellowhammers, and a few Corn Buntings make use of the margins. Skylarks, often nesting in 'Skylark plots', scour the ground for insects to feed on. From slightly higher vantage points John shows me some of these small Skylark plots amongst the crops; the giveaway is the almost continuous singing of Skylarks way above us, a heartening sound given that our visit is not in the best of summer weather.

Countryside Stewardship in England pays £532 per hectare for field margins such as those on the Overbury Estate and recommends that they be up to six metres wide. The payment is calculated to compensate for the absence of a saleable farm crop from the land. Topping the growth at a height of around 30 cm to prevent seeding of undesirable weed species such as Wild Oats and Creeping Thistle is allowed under the scheme during the growing season because many of the desirable flowering arable plant species are shorter than this. Where a pernicious weed burden develops over more than 40% of the area, targeted broad-spectrum herbicides can be used once annual species have set seed (normally in September). Therefore, creating such flower-rich margins, thereby developing an important wildlife habitat on a lowland arable farm, shouldn't present any major problems for most farmers.

Whilst it's natural to assume that taking land out of crop production to provide crop margins full of flowers and grasses will deplete crop yields on a farm (see Chapter 5), the opposite is true – crop yields are boosted, making up for the loss of crop from the flower margins! It's a research finding that needs to be considered when the compensation element of these agri-environment measures is reviewed.

Another concern many farmers express if they are contemplating setting aside some land to grow native flowering plants alongside farm crops is the possibility (many imagine a probability) that pernicious, perennial plants, such as docks, thistles, ragwort, and nettles, will move into the farm crop. Intensive farming encourages plants like these because they often favour well-fertilised soils; most arable soils are very well fertilised.

But according to one major study, the notion that flower-rich margins on arable land can lead to 'weeds' infiltrating into the crop is overstated.[i] It found no relationship between how such field margins were managed and the abundance of weeds in the adjacent crop. Margins established by sowing a wildflower seed mix had very few perennial or annual weeds in them for the first few years. With time, such weeds increased, although their presence could be controlled by mowing the margin. This study also found that structural diversity – a mix of different wild plants of varying growth heights – is as important in these margins as the diversity of plant species. Tussock-forming perennial grasses, for instance, provide habitat for overwintering invertebrate predators such as carabid beetles and spiders.

At Newlands Farm, 300 hectares of rolling land on the south coast of Dorset, Paul Simpson is retaining a different set of plants. Alongside a sandy track he shows me a swathe several metres wide of what most farmers would most definitely call weeds growing adjacent to a large, intensively-managed field of oilseed rape just about to burst into vibrant yellow flower. This vegetation is a pollen-rich mix of common wild plants sown to attract bumblebees, butterflies, and hoverflies in summer. There are clovers, Red Campion already in flower, and I spot some Musk Mallow which will flower pink in midsummer. It will be good habitat, too, for a plethora of mice and other small rodents as well as for seed-eating birds when many of these flowers set seed.

Mr Simpson has an Environmental Stewardship agreement which includes nearly two hectares of such wild plant mixes on different parts of his farm. He's also leaving four hectares of unharvested and unsprayed margins around barley fields to encourage once common cereal field 'weeds' to grow up, plants like those Jake Freestone is encouraging in his arable margins on the Overbury Estate. Altogether, Paul Simpson has included over 16 hectares of his farm in an Environmental Stewardship agreement. That doesn't include another 30 hectares of spring barley he grows so that he can retain the crop's stubble overwinter for seed-eating small birds like Yellowhammers, Goldfinches, and much rarer Corn Buntings.

Not being able to control pernicious weeds with herbicides, either because some have become resistant or because there are human health concerns surrounding their continued use, is worrying many farmers. On farms the most commonly used wide-spectrum herbicide (it kills virtually all plants) is glyphosate. It's frequently used by gardeners too.

EU reviews of pesticide regulations are threatening the continued use of glyphosate because traces can remain on food crops, something first noticed in

Keeping it wild. Culm grassland in Devon on which Don Osborne grazes Highland cattle.

2015 when the International Agency for Research on Cancer (the specialist cancer agency of the World Health Organisation) said that its active ingredient was 'probably carcinogenic in humans'.[ii] It remains licensed for use in the EU but only until the end of 2017, by which time the European Agency for Chemical Products should have re-assessed its safety and a decision will have to be taken to ban it, restrict its application, or allow its full use (in December 2017 the EU re-licensed glyphosate for a further five years). A large number of other pesticides are also under review, much to the irritation of most farmers who rely heavily on their use to control pernicious weeds.

A handful of lowland farms have substantial areas of valuable surviving wildlife habitat that have not been destroyed by agricultural intensification. Now in an Environmental Stewardship scheme, Leasefield Farm near Beaworthy in North Devon has 40 hectares of land which is so wet and knee-deep in rank grasses and other vegetation that the owners, Rupert Weatherall and Don Osborne, considered it a 'write-off' agriculturally when they bought the farm seven or eight years ago.

'My God, what a mess! That's what we thought when we bought this place and we looked at the wet, scrubby grassland that dominated all the bottom part of the farm. We thought we'd sell it on and buy something better. The whole farm isn't much more than 100 hectares so we were very skeptical it would be any good to us at all,' says Mr Weatherall. 'But how wrong we were! We've had sound advice from the Devon Wildlife Trust and Environmental Stewardship money to manage it and it's become a valuable part of the farm. We've taken a lot of scrub out and the cattle, some of them Highlands, very hardy beef cattle, do extremely well on it.'

To botanists, the land Mr Weatherall is talking about is a very uncommon habitat rich in an array of damp grassland plants and a wealth of insects. Culm grasslands (so-named after the geological formations underlying them) have developed on low-lying, acidic clay soils that are poorly drained. Found only in this part of North Devon, South Wales, and south-west Scotland, some

92% of it has been lost in the last century because of agricultural drainage and ploughing, planting it up for forestry, or letting it scrub over with willow and other trees.[iii]

I'm walking over the land with Don Osborne and Lisa Schneidau, the Devon Wildlife Trust's Project Manager for their Northern Devon Nature Improvement Area. Covered in tussocks of rushes and pale Purple Moor Grass, it is certainly very boggy; it's not at all easy to make progress. Patches of willow and birch scrub sprout from the wet grassland here and there; much of it has been cleared by cutting, and the Highland Cattle keep a lot of the regrowth in check. Marsh Thistle is abundant, its narrow green leaves dotted throughout the grass tussocks although we are too early in the season to see its glorious purple flowers. However, pink-coloured Heath Spotted Orchids are in flower and Lisa tells me that soon the whole place will be specked with thousands of white Lesser Butterfly Orchids and tall, rich purple-coloured Southern Marsh Orchids. She says it's a sight to behold, so I wish I had visited a little later in the season!

There are many other notable plants here too: the uncommon frothy-white blossoms of Whorled Caraway, the yellow-flowered St John's Wort, and blue flowers of Devil's-bit Scabious. Where the soil is particularly waterlogged, bog vegetation comprising sphagnum mosses, Bog Pondweed, and sedges has developed. The huge numbers and variety of plant species support a rich and diverse community of invertebrates too: moths, butterflies, dragonflies, and many more as well as a range of small mammals and reptiles. And there are multitudes of breeding birds including hunting Barn Owls, Woodcock, and Grasshopper Warblers with their monotonous, insect-like reeling 'songs'.

Places like these aren't just valuable for the plants and other wildlife they nurture. Research by the Devon Wildlife Trust (DWT), the University of Exeter, the Environment Agency, and others has shown that culm grassland can hold up to five times as much water as a closely grazed grassland in extreme weather events, thereby holding back flood waters.[iv] It also locks up considerable amounts of carbon, most of it released as carbon dioxide, a potent greenhouse gas, if such grassland is drained.

The North Devon Nature Improvement Area covers 72,000 hectares of North Devon, most of it in the catchment of the River Torridge. The scheme started in 2012 as a partnership that included the DWT, Devon County Council, Butterfly Conservation, the Woodland Trust, and many others; by 2015 it had restored nearly 1,500 hectares of culm and other plant-rich grasslands. 'Some grasslands (often subject to ploughing up and reseeding in the past) are on their way to becoming more diverse in flowering plants because we have helped farmers spread green hay or seed taken off nearby meadows as a hay crop,' says Lisa. In all, the scheme's advice to farmers has set up 56 different Environmental Stewardship agreements.

The three of us walk over a three hectare pasture on a gentle, well-drained slope leading down to Mr Osborne's culm grassland; it was a regularly-manured ryegrass pasture until three years ago. 'We spread some green hay on here taken from a flower-rich meadow and we're starting to get more flowers in it,' he comments. 'We're getting back buttercups, a bit of Hay Rattle, some Lady's Smock and forget-me-nots; not very special yet but we're hoping more flowers will come in. We don't use any fertiliser here, just cut it for hay midsummer and put cattle on to graze it until spring,' he adds.

'We have been inundated with enquiries from farmers about the Area Improvement Scheme,' says Ms Schneidau. 'Some of the attraction is the Stewardship payments for land they can't always make a lot of use of. The existing scheme works well but I have a concern that the new Countryside Stewardship scheme might require too much detail from farmers and be unnecessarily bureaucratic. Farmers need advice to draw up their proposals; the scheme doesn't pay for that and we have to make a charge for the advice we give them.'

Good, practical advice and assistance have been essential for these farmers to recreate habitats and manage existing habitats to encourage more wildlife on their farms. A multitude of organisations offer advice, both from the voluntary and the commercial sector, so it can be somewhat confusing to know who to turn to if you are a farmer wanting to better cater for wildlife on your land but unsure how to go about it. Perhaps Farm Wildlife, a consortium of eight voluntary organisations including the RSPB, The Wildlife Trusts, and smaller outfits such as the Bumblebee Conservation Trust, is therefore a breath of fresh air!

Its website claims that it's been developed with farmers for farmers, bringing together best practice advice from all of their partner organisations to help farmers do the best for wildlife.[v] It gives general guidance about how to construct a plan for a farm, how to work out what habitats to retain and/or provide and where, at what scale, and how they need to be managed alongside a farm's crops or livestock. It also gives a good number of practical examples of what has been achieved by a wide range of farmers.

Although it promotes a 'Find local advice' tab on its website, many regions in the UK still have no representative to contact, and it doesn't provide advisers who can give integrated advice to farmers. Instead, it provides contact details for representatives of individual NGOs knowledgeable about their own specialist area, be that birds, butterflies, or plants.

Kathryn Smith, Farm Wildlife's Project Manager, acknowledges that the service is in its infancy (in July 2016) and accepts the shortcoming. 'All of these advisers will be able to provide advice that echoes the ethos of the content of the website, or more specialist advice where required for rare or sensitive species. Again, as the project develops, we may find ourselves in a position to provide a wider range of appropriate contacts. This will be something that evolves over time if we decide to take that approach,' she says.

It's a similar 'service' to that offered by FWAG, the Farming & Wildlife Advisory Group, established in the 1960s by a group of farmers who considered wildlife as an important part of farming. But it operates only in parts of southern England and the Midlands. Small, independent FWAGs operate in Wales and Scotland.

Farm Wildlife has three plans for flower-rich habitats on arable land. The first is to provide the right conditions to bring back arable flowers on lowland farms that are growing cereals and other crops by creating field margins or headlands that are unsprayed and unfertilised but are ploughed annually. That's what Jake Freestone has done on the Overbury Estate. The second is to create small areas of habitat rich in native wildflowers, in field corners for instance, by leaving a wide grassy margin at the side of fields, against a hedge maybe, unsprayed, unfertilised, and cut just once a year in late summer to stop it being taken over by scrub, rather like some of the work done by Paul Simpson in Dorset. The third, albeit more expensive, is to sow a flower mix using species that are known to produce plenty of pollen and nectar to attract a wide array of insects. They might be grown in strips alongside a farm crop, again like Mr Simpson has done at Newlands Farm.

Few farmers, though, have gone to such lengths to retain and recreate so many flower-rich hay meadows as Tom Lord at Lower Winskill Farm in the Yorkshire Dales (see Chapter 9). His 14 hectares of upland hay meadow are traditionally managed: manured but without any synthetic fertilisers or pesticides, and cut after the flowering plants have set seed in late summer. But when Mr Lord took over the farm 30 years ago, many of the meadows were full of pernicious thistles and little else was visible. By cutting the fields year after year before the thistles flowered he slowly encouraged a wealth of flowers and grasses to become more dominant while the thistles were gradually depleted. The agri-environment scheme run by Natural England helps to fund their continued management. Now in summer, Yellow Rattle, Common Spotted Orchid, and Harebells are just a few of the flowers that thrive in these upland, stone-wall bordered hay meadows. They are a sight to behold.

Agri-environment measures designed to bolster populations of invertebrates, small mammals, and birds can often deliver significant benefits for plants too. That should be no surprise; both small and larger species of animals are not going to fare well if they don't have the necessary vegetation to survive and flourish. A good example is a study that investigated the spinoff for plants and invertebrates of measures designed to boost Corncrake numbers in the coastal machair vegetation (sandy, flower-rich grasslands) farmed on some Scottish west coast islands. With very few exceptions it found that flowering plant species' richness and the abundance of late-season flowers, butterflies, bumblebees, and the majority of invertebrates all benefitted.[vi]

Sowing arable crops in spring rather than in autumn, thereby leaving stubble through the winter, is known to be valuable for insect and seed-eating small birds (see Chapter 7), and it's also beneficial for wild plants. It allows a range of fast establishing 'weed' species to grow through the autumn, some of which will flower and produce seeds either before winter sets in or the following spring before a new cereal crop is sown.

THE MEADOWS THAT FADED AWAY

There were once wildflower meadows like those at Lower Winskill in almost every parish the length and breadth of Britain, but today only 2% of the meadows that existed even in the 1930s remain. Three million hectares of wildflower meadow have been lost so far and they are still being destroyed.[vii] What had been a widespread and ubiquitous part of farming and people's daily lives disappeared almost entirely in the space of a single generation. Huge areas of grassland were ploughed to grow cereals during World War II, and this started a process which would see lowland meadows decline to mere remnants in the following 40 years.

Those that survived are often 'one-offs', remaining in isolation surrounded by a sea of more intensively-managed farmland which prevents many of their wind, bird, and insect-dispersed pollen and seeds from establishing in adjacent fields. Instigating the correct protection and farming management of such remnants is of little use unless the remaining examples can be connected up by converting the land in between – maybe across neighbouring farms – back to hay meadow management. With willing farmers it can be done passively by leaving natural seed dispersal to take its course and gradually re-establish the flowering plants that have long been absent. It can also be done more actively, as Henry Edmunds has done on parts of his Cholderton Estate or Don Osborne at Leasefield Farm, by harvesting seeds from part of an existing meadow and spreading it on the land to be reverted.

At Honeydale Farm near Stow-on-the-Wold in Gloucestershire, Ian Wilkinson – the owner of nearby Cotswold Seeds, seed supplier to many farmers and growers across the UK – is experimenting with different crop seed mixes, so-called herbal leys, both to produce nutritious grazing for sheep or cattle and to provide a greater variety of plant species, thereby increasing its wildlife value. Most livestock farmers in the lowlands graze their dairy cattle or sheep on temporary grasslands (leys) sown mainly with cultivated varieties of ryegrass and clover, the clover used as a nitrogen-fixing crop. Being so poor in plant species, this sort of grassland is of virtually no wildlife value. Leys require large amounts of fertiliser to keep them productive and they are often ploughed up and re-sown every few years to keep them at peak productivity. Such high productivity means that their grass and clover growth can be cut maybe three or four times in a season to make silage for winter cattle feed. Although the seed costs of a more diverse herbal ley are maybe 25% more than a ryegrass-clover ley, because herbal lays need no artificial fertiliser and because they provide more nutritious forage for livestock, thus improving animal health and lowering vet bills, the overall costs are less.

Leys have replaced most permanent pastures and hay meadows in the British lowlands. This changeover has been one of the causes of the greatest loss of flowering and other plants on our farmland in the last few decades. These losses are not only of plants but of a plethora of different species of invertebrates, small mammals, and birds. Around 97% of traditional hay meadows disappeared between the 1930s and 1984.[viii] What Ian Wilkinson is showing is that there's a middle way.

Cotswold Seeds claim that the herbal leys at Honeydale Farm can support grazing early in the year and continue to produce forage right through the summer and autumn. They can consist of a mix of grasses including some ryegrass, Cocksfoot, Timothy-grass, and Meadow Fescue, plus legumes such as Red Clover, White Clover, Alsike Clover, Birdsfoot Trefoil, and deep-rooting forage herbs including chicory, Ribgrass, Great Burnet, and Sheep's Parsley. Different mixes of plant seed are available for different locations and soil or climate characteristics. A large amount of Sainfoin can also be included, and when these are all combined a herbal ley not only improves soil structure but also draws up essential vitamins and minerals from deeper down in the soil for the ruminant animal.

Ian Wilkinson's farm staff are using herbal leys for so-called 'mob grazing' by sheep, sectioning off different areas of a large field with electric fencing and allowing sheep into one section to graze it down before moving them on to another section while the previous area regrows. Allowed to grow tall and flower in a narrow strip at one end of the field, the herbal ley I looked at provided a colourful mix in which the bright blue flowers of chicory on its tall stems were competing with butter-yellow Birdsfoot Trefoil lower down in the vegetation and the big conical heads of strikingly pink-flowered Sainfoin.

'Sainfoin grows well in drought conditions – it has a long tap root – a factor that might become increasingly important because of climate warming, particularly in the south of Britain,' says Mr Wilkinson. 'There's an increasing interest from many lowland farmers in growing it as a crop mixed with sown grasses for silage making and in these times of rising fertiliser, feed, and veterinary drug costs, it's a crop that can be of huge benefit to farmers and the environment. On the right calcareous soil like here at Honeydale, Sainfoin will last up to five years or more and will thrive in a wide range of temperatures from the very hot to the very cold.'

'I'm growing a great deal of Sainfoin among my grass crops. It's an important flower for bees,' comments Henry Edmunds on an early June day as he shows me some well grown fields on his Cholderton Estate near Salisbury, where the plant's pink flowers are starting to show. 'We could cut it for hay but we make silage from it. It makes a very good winter cattle feed. After cutting it, we graze the fields with sheep or cattle until the following spring when we close the fields up to let the crop grow again. We don't add any fertilisers; just some cattle manure, and certainly no pesticides. Most farmers around here would plough this lot up and grow pesticide-drenched wheat or barley. Sacrilege! We have 120 hectares here of grassland with Sainfoin sown in it,' he adds.

AS PRETTY AS A PEA

Sainfoin (it means 'wholesome hay') is both a native plant in the UK and also a forage crop that has been grown in many parts of the world, Europe included, for hundreds of years. It was grown for cattle and sheep grazing and for hay making.[ix] Native to South and Central Asia, it was introduced into Europe in the 15th century. For farmers with dry, calcareous soils Sainfoin is an ideal crop. It has a tap root that grows down to a great depth, making it highly resistant to drought, and its roots are able to draw up minerals from well below the topsoil. In the days when working horses were commonly used on farms, Sainfoin provided them with high-quality forage. Traditionally grown as part of a three to four year rotation, it was also used to add fertility to the soil before arable cropping. In the south of England, until Victorian times, one in seven fields used to be covered with Sainfoin.

A sturdy, upright, perennial legume of the pea family, with striking heads of bright pink flowers, Sainfoin is particularly nutritious for livestock. It also contains tannins which provide it with an anthelmintic property to naturally reduce intestinal parasitic worms in sheep and cattle. With many such diseases becoming more common, and concerns about antibiotic resistance growing, any naturally occurring drugs need to be examined thoroughly.

Flowering from June until September, it attracts both bumblebees and honeybees – ten times as many as White Clover – and it is commonly stipulated in agri-environment 'pollen and nectar' mixes for headlands in arable farming to provide a great food source for insects. When Sainfoin flowers, it is so appealing to honeybees that they will often ignore all other nectar sources to forage on it.

Recent research has confirmed that small improvements in the plant diversity of pastures can attract more pollinating insects. Common flowers such as dandelions and Creeping Thistle – both much despised by farmers and of no agricultural value – have very high levels of visitation by a wide range of insect pollinators. Introducing plants such as chicory, a flowering plant that attracts a variety of pollinators and is of agronomic value (especially for sheep and lambs), with proven anthelminthic properties and deep tap roots that cope well with drought, can provide substantial agricultural and wildlife gains. The researchers found that a modest increase in the diversity of plants encouraged to grow in agricultural grasslands – legumes and other flowering plants – which is achievable with the expertise and resources available to most grassland farmers, could enhance the numbers of pollinating insects and bring advantages for the pollination of adjacent farm crops too.

Planting trees in pastures, unless they are confined to hedge lines, isn't something that most farmers would exactly welcome with open arms. After all, a scattering of sizeable oaks or ash in the middle of a field gets in the way of most farming operations and takes huge quantities of water and soil nutrients away from whatever crop is being grown there. But research backed by the Woodland Trust is starting to show that there are untapped benefits to having some individual trees dotted here and there on farms, for the farmer as well as for wildlife.[x] A mature oak tree, for example, can provide a habitat for hundreds of species of plants, invertebrates and other animals and planting shelter belts of trees on hill livestock farms can improve animal welfare and productivity. One convert the Woodland Trust quotes is Jonathan

Good for insects; good for livestock: Sainfoin is gaining in popularity as a component of herbal leys (courtesy of Javier Martin).

Francis, who farms sheep and cattle near Caersws in Powys. He's planted well over 2,000 tree saplings with help from the Trust. 'Shelter is a big issue for us, and all our rain seems to come at once these days,' he says. 'We're already seeing less waterlogging in the ground and I'm hoping we can turn our animals out of the sheds earlier in spring, saving us money.'

Research also supported by the Woodland Trust is suggesting that cattle able to browse nutritious fodder, including tree leaves, are healthier than those that aren't.[x] Dr Jim Waterson, Head of Forestry at Harper Adams University where the research is based, says: 'Willow bark is an active ingredient in aspirin and its anti-inflammatory properties might help cows with mastitis or sore hooves. Minerals in tree fodder could also reduce gut worms or even aid resistance to bovine TB. Our findings could unlock big savings in antibiotics and food supplements as well as driving the case for trees on farms.' The Woodland Trust's link with Harper Adams has also investigated how trees on arable land boost soil earthworm populations. Rather than the accepted wisdom that trees shade crops and reduce yields, their leaf litter enriches the soil while the trees shelter the crop from wind.

Getting more plant diversity on farmland – thereby boosting invertebrate, small mammal, and bird populations into the bargain – isn't especially difficult.

Advice is available to farmers from a number of organisations, including national and local conservation bodies, farming and wildlife advisory groups, and commercial organisations. No farmer anywhere in the UK can fail to create or better manage existing wildlife habitat because he or she can't obtain advice.

Financial incentives are available through the agri-environment schemes run under different names in the four UK countries. However, it's clear that since most of these schemes are targeted at certain areas of the countryside, they are generally competitive (so farmers offering greater wildlife benefits are more likely to be accepted), and their budgets are limited compared to the current budgets for the Basic Payment Scheme (BPS) subsidies, many farmers will be unable to secure an agreement. There is also a concern expressed by some farmers that the schemes as they are presently constituted are too rigid and don't allow enough flexibility for a participating farmer to modify the timing or nature of his operations to take account of factors outside his control, the weather being a very obvious one.

Notwithstanding grumbles that are fair or otherwise, limited budgets, and the resulting restrictions in the number of farmers that can qualify for agri-environment agreements, these schemes are the only current, widely available mechanism to bolster the wildlife content of our farmland. Where they have been implemented, the farming examples quoted in this chapter show that there have been enormous gains in turning around the huge losses of farmland flowers and other wild plants that have been caused in particular by farm intensification through the last few decades. Post CAP, it is vital that more cash, directed away from BPS subsidies currently paid to profitable lowland farms, is invested instead in the agri-environment schemes the four UK governments adopt in the future. The survival of many of our farmland plants and animals depends on it.

Endnotes

i   'From weed reservoir to wildlife resource – redefining arable weed margins,' by Helen Smith et al. *Wildlife Conservation on Farmland, Volume 1*. Edited by David Macdonald & Ruth Feber. Oxford University Press, 2015.

ii   http://www.iarc.fr/en/media-centre/iarcnews/pdf/MonographVolume112.pdf.

iii   www.northerndevonnia.org.

iv   'The Economic Value of Ecosystem Services Provided by Culm Grasslands,' Charles Cowap et al. Devon Wildlife Trust, 2015.

v   https://farmwildlife.info/.

vi   'Agri-environment management for Corncrake delivers higher species richness and abundance across other taxonomic groups,' N. I. Wilkinson et al, 2012. *Agriculture, Ecosystems and Environ ment*, 155:27-34.

vii   'Save our magnificent meadows,' Plantlife, 2010.

viii   'The changing extent and conservation interest of lowland grasslands in England and Wales: A review of grassland surveys 1930-1984,' Rob Fuller, 1987. *Biological Conservation*, 40: 281-300.

ix   'Introduction to Sainfoin,' LegumePlus.eu.

x   'Modest enhancements to conventional grassland diversity improve the provision of pollination servic es,' Katherine A. Orford et al, 2016. *Journal of Applied Ecology* doi: 10.1111/1365-2664.12608.

CHAPTER 7

Putting up the Bunting

There is nothing in which the birds differ more from man than the way in which they can build and yet leave a landscape as it was before.

Robert Lynd (1879–1949), Irish writer, poetry editor, and literary essayist

On a cool, damp, and slightly drizzly June morning, the view out over Labrador Bay on Devon's south coast near Teignmouth is not quite as spectacular as it can be. But it's here on the steeply sloping coastal fields down to the Bay that what is probably the most spectacular farm bird recovery project in Britain has quietly succeeded.

Almost as soon as I arrive in the small car park perched high above the Bay, I spot a pair of rusty brown Cirl Buntings, the male being a giveaway with his yellow and black striped head. They are flitting from bush to bush and fencepost to fencepost in a wide ribbon of scrub and hedge not 50 metres downslope. Cath Jeffs – the Cirl Bunting Project Manager for the RSPB in Devon – tells me that in 1989 there were just 118 pairs of these little seed-eating buntings left in Britain, all of them along the south coast of Devon. Once so common around the villages and farms hereabouts that it was known locally as the Village Bunting, changes in farming practices began to steadily deplete their numbers. However, a count of these gorgeous, noisy little birds in 2016 found over 1,000 pairs, the target the RSPB had set itself to reach by 2020!

The RSPB owns 52 hectares of land here above Labrador Bay, a coastal strip about two kilometres wide where the farming is done by the RSPB's tenant, Peter French. It's Mr French's mode of farming, worked out between him and Ms Jeffs, that is the key to the huge success of this project.

Agri-environment success. The Cirl Bunting in Devon (courtesy of Paco Gomez).

Peter French farms nearly 300 hectares of land here, including the part he tenants from the RSPB. 'It's been in my family for four generations and I'm growing wheat, barley, oats, oilseed rape, and maize on the most productive land,' he says. 'Here on the RSPB land I grow some fields of barley, but it's spring sown so we leave the stubble on the land all winter after harvesting the crop. And the pastures, they're pretty steep, but we have beef cattle grazing them and sometimes sheep. Most of the spring barley fields are very low input so we apply hardly any fertiliser and no pesticides.

'It's all part of the Environmental Stewardship agreement I have with Natural England so I get payments annually to help me manage the land this way. I'm perfectly happy with it. The yield from the spring barley is half what it is on my intensively-managed fields, two tonnes per hectare, not 4.6 tonnes. But once I've costed in the fertiliser and pesticides I use on the intensive barley, the gross margins aren't much different!'

On his intensively-managed land not in the Environmental Stewardship (ES) scheme, Peter French sprays his crops to control weeds such as cleavers, chickweed, groundsel, and wild oats, and uses insecticides to kill off aphids as well as using fungicides to control infections such as mildew on his wheat and barley. This is a part of Britain with a high rainfall so the frequently damp conditions encourage fungal infections like mildew. What are considered 'weeds' in the farm crops are, of course, native British plants (see Chapter 6), all of which set seed in late summer and are high on the menu for Cirl Buntings and other seed-eating birds such as Linnets. He has planted up one small field with a bird seed mix – millet, barley, and quinoa seed – another useful measure included specifically in his agri-environment scheme to provide more food, especially in winter.

Although the spring barley is vital for the Cirl Buntings, it has been less common since the change from the traditional spring sowing of cereal crops to autumn sowing put paid to most stubble fields from the 1980s onwards. After the barley is harvested in summer the stubble provides plenty of seeds that develop on the scatter of flowering plants left behind. A few grains of spilt barley the combine harvester might have missed are soon snapped up too. Without these seeds, the Cirl Buntings might not be able to survive through winter. Across the country, such crop stubbles supported not only Cirl Buntings but also Yellowhammers, Tree Sparrows, Linnets, Goldfinches, and many more. The loss of these stubbles has depleted the numbers of many once-common farmland birds. In the last few years though, farmers have started retaining winter crop stubbles again as a means of complying with tightened EU regulations for receiving their BPS subsidies, for complying with the EU Water Framework Directive (see Chapter 8) and, in part, the current Ecological Focus Area requirements of 'greening' (see Chapter 4). As it becomes a more widespread practice, it's likely to help the populations of seed-eating birds to recover.

Through the growing season, on the edges of the barley fields where Peter French's crop thins out, there are patches of bare soil with native plants growing up, places that Cirl Buntings will find plenty of small insects to feed their chicks on. If the crop was sprayed with an insecticide, there would be virtually none. The grazed but unfertilised fields of permanent pasture here are a rich source of grasshoppers and other insects that the buntings feed their nestlings on too. Since these pastures are never ploughed or fertilised, they retain a mix of grasses and other flowering plants which support healthy populations of invertebrates.

Yet Cirl Buntings need breeding sites, not just feeding fields; that's where hedges come in. Leaving hedges to grow wide, tall, and dense; replanting some that have previously been grubbed out; and only trimming them every few years so they are allowed to ramble provides a thick cover of shrubs and a welter of flowering plants either side of them. That's just what the buntings need to nest in; they also make good use of patches of scrub Peter has left here and there on his cattle-grazed pastures. Thick hedges and scrub patches aren't just good for buntings and other birds; they are useful shelter from coastal gales for the livestock too.

'This is the ideal combination of breeding and feeding habitat for the buntings,' comments Cath Jeffs, as we watch three male Cirl Buntings line up very close to us on some hedgerow branches while a couple of the slightly less flamboyant females scour the edge of a nearby barley crop. 'Just on our reserve here, we now have 21 pairs breeding; in 2008 we had three pairs left. We're also advising other farmers along the South Devon coast to encourage them to provide this mix of habitats on parts of their farms. Advice is the key; we've had an adviser here to work with farmers since 1994. Some have entered the Stewardship scheme, others are just taking our advice; in general I find that they respond well and support what we're trying to do here,' she adds. The Cirl Bunting project has been running for 25 years, and has given advice on 12,000 hectares of farmland involving collaboration with well over 300 farmers.

Nirvana for breeding farmland birds. Well-grown hedges alongside pesticide-free barley.

'Apart from the buntings we have good populations of Linnets, Bullfinches, Stonechats and Skylarks,' comments Cath Jeffs. 'We used to have Grey Partridge but they've gone and Turtle Doves are now a rare sight.' Advisory work which started in the early 1990s to help restore Cirl Buntings probably came too late to help them.

Other individual farmers are foregoing productivity and making use of an agri-environment scheme in order to help return once-familiar farmland birds too. On Henry Edmunds' 1,000 hectare Cholderton Estate, the emphases on his farmed land are sustainable organic farming; balancing the demands of modern highly-competitive agriculture and the preservation of the countryside; and protecting the best parts of his Estate for wildlife, including breeding Lapwings.

Jumping out of his battered Land Rover, we stop alongside a barley field, spring sown in order to leave the stubbles over winter. No pesticides – and no fertiliser except manure – are applied to this field; Charlock (Wild Mustard) had grown up in the centre.

'It's the strong odour of the Charlock that deters foxes,' says Mr Edmunds. 'The Lapwing chicks feed on insects in the spring barley after they leave their nests in the middle of the field when the Charlock starts to grow up. Just in case the foxes aren't put off, we have an electric fence around the outside of the whole field too. We'll harvest the barley later in the summer when the Lapwings have moved on, many of them into an adjacent field which is also protected with an electric fence. In there we have some sheep grazing the well grown pasture and the young Lapwings like it for feeding. Most farmers are using too many pesticide sprays; their crops are sprayed to hell,' he adds.

THE BUNTING THAT'S UP

At the end of the 19th century, Cirl Buntings were commonplace across southern England, with smaller numbers in central England and parts of Wales. Their distribution and numbers started to decline around the 1930s but crashed in the 1960s and 1970s, almost certainly due to farmers changing over from spring to autumn sowing of cereals, thereby removing the stubbles that supported them through winter.[i] The trend of increasing field sizes by grubbing out lots of hedges – therefore taking away vital breeding ground – was a major factor too. Farmers specialising into either pasture for livestock or arable land to produce high-yielding crops has also meant a reduction in mixed farming. That means that arable crops and pasture have been physically separated; bad news for rather sedentary Cirl Buntings that need both close at hand.

A successful re-introduction on the Roseland Peninsula on the south Cornish coast (using nestlings from Devon) confirms that changes in farming practice – funded at present by Countryside Stewardship schemes – might well succeed in returning this bird to even more of its former breeding range in the UK. The first successful breeding in Cornwall took place in 2007 and there are now over 50 breeding pairs. 'We've worked with local farmers, the National Trust, and other conservation NGOs to increase the amount of suitable habitat for the birds, and we have a farmland adviser working with landowners to secure further habitat for the wider, natural spread of birds using Natural England's agri-environment scheme,' says Cath Jeffs.

This mosaic of grass and spring tillage fields close together has declined significantly on most farms in recent years, and the loss has resulted in a decline in Lapwing numbers. They were still common breeding birds on the Estate in the 1960s, although most had been lost by the late 1980s; Cholderton had only three breeding pairs left. In 2016 though there were around ten pairs of Lapwings breeding on the Estate, and some years there are more.

Their decline nationwide actually began much earlier, probably late in the 19th century, as a result of commercial egg collecting for food and the increasing drainage of wet lowland fields and marshes, as well as drainage and ploughing of land in the uplands.[i] In recent decades the decline has accelerated with more intensive management of farmland, particularly drainage of wet pasture; the change from hay cutting to silage making; ploughing up of permanent pastures to grow arable crops or to replace them with temporary, heavily-fertilised pasture (grass leys); the changeover to autumn-sown cereals; and increased livestock numbers trampling nests and preventing the growth of areas of rank grassland. UK-wide, breeding numbers plummeted by 63% between 1970 and 2014, with the greatest losses in southern England and Wales.[ii] With funding from England's agri-environment scheme to compensate for any productivity losses, Henry Edmunds is doing precisely what is needed to bring Lapwings back while still producing farm products – barley and sheep – albeit at lower yields than he could if the land was managed intensively.

Elsewhere on the farm, Mr Edmunds is supporting Corn Buntings by leaving areas of rank grassland, growing spring-sown barley without pesticides, and allowing patches of scattered scrub to develop to provide breeding sites. So far just a few pairs breed here, though a large number winter on parts of the Estate, encouraged by stubbles. Corn Bunting numbers have been dropping for decades, perhaps even before 1900, as the area of cereal growing has declined.[i] Also, more and more areas have become intensively-grown with the help of pesticides, autumn sowing has replaced spring sowing (depriving Corn Buntings of winter food), and hedges – used as breeding sites – have been grubbed out in more recent decades. Between 1970 and 2013, their numbers fell by 91% overall in the UK, and they are still falling today.[iii]

Skylarks – which do well here in the spring barley fields – and Grey Partridges are thriving amongst Mr Edmunds' crops. These two farmland birds have declined enormously UK-wide because of the intensification of farming in recent decades. Other notable birds on the Estate include the Barn Owl, Hobby, Long-eared Owl, and Bullfinch. The Estate has erected almost 30 owl boxes in barns in different locations and the number of Barn Owls has built up over several years from a single breeding pair to about seven.

To bolster Grey Partridge numbers on the Cholderton Estate, Henry Edmunds has increased legal fox and crow control, maintained a mixed farming system to provide a diversity of cropping and grassland, increased the area of spring-sown barley, protected hedgerows from animal grazing, converted to a fully organic system, and kept unfarmed patches of rank grassland all over the Estate to provide shelter and nesting cover. But many farmers should be able to bolster Grey Partridge numbers with fewer changes to their farming practices. The important measures are: to take care not to use insecticides unless absolutely necessary, leave several metres unsprayed around the edges of fields, plough stubbles into the soil as late as possible, and create permanent and rarely-cut grassy margins around arable fields. Fencing off pasture edges so that they can't be grazed but can be cut on rotation around the farm every few years helps too.

On the spectacular south coast of Dorset near West Lulworth, Paul Simpson farms around 300 hectares of mostly permanent grassland and flower-rich downland on which he raises beef cattle and sheep. On just over 50 hectares he grows wheat, barley, and oilseed rape on rotation; the barley is spring sown. While Grey Partridges have almost certainly always been on the farm, it wasn't until 2003 that they were first recorded systematically. Since then they have increased, although not steadily; in some years the autumn counts done by the Game and Wildfowl Conservation Trust (GWCT) have fallen.

DOWN AND DOWN

The decline of the Grey Partridge almost exactly matches that of the Corn Bunting but is even more depressing: their numbers fell by a staggering 91% between 1967 and 2010 UK-wide.[iv]

The causes include the demise of tussocky grassy areas such as those around the base of hedges – their breeding sites – and a loss of invertebrates from these areas. Another is the growth in herbicide use over the last few decades, killing off 'weeds' that provide invertebrate habitat amongst farm crops and around their edges. Invertebrates are especially important for rearing partridge chicks because they need a protein-rich diet to flourish. The elimination of winter cereal stubbles has also aided their steep decline. As the numbers have dropped, the relative impact of foxes and crows killing the chicks has increased. Infection with intestinal nematodes picked up from farm-reared pheasants may be contributing to the decline as well, while the practice of releasing large numbers of Red-legged Partridges for shooting (a bird not native to the UK) can lead to Grey Partridge extinction, in part because shooters are unable to readily distinguish the two species.

But the general pattern on the farm is an increase – a peak autumn count of 35 birds in 2011 though more like 20 more recently. What is more heartening is that the ratio of young to old birds is frequently around three or four young to old; according to the GWTC, anything above 1.65 is a sign that there are enough birds to replace adults lost naturally or to predators.

'I've had a Stewardship agreement since 1993 covering hedge restoration and overwintered barley stubbles but I've added in other things over the years: bird food plots where I sow seed mixes including millet, linseed, and mustard, and beetle banks as well as permissive paths for people to walk over the grasslands and downs,' says Mr Simpson. Beetle banks are raised earth ridges with dense tussocky cover providing warm and dry areas for invertebrates and farmland birds, especially for nesting Grey Partridges.

Much of this provision of habitat has helped a tentative recovery of Grey Partridges at Newland's Farm, bucking the national trend. UK-wide their numbers fell by 92% between 1970 and 2014 though there has been a very recent slight increase.[ii] His modified farming practices are paying dividends, and not only for Grey Partridges.

As we drive around his farm, Mr Simpson tells me that the land supports decent numbers of Brown Hares, Stonechats, plenty of Skylarks, Yellowhammers, and a few Corn Buntings. Moreover, there are a good array of butterflies on the flower-rich pasture and downland, about 50 hectares or more of which is designated as a Site of Special Scientific Interest (SSSI).

Getting wildlife back on most farms is not rocket science, but three factors have to be aligned. Firstly, the farmer has to show an interest in helping to conserve wildlife on his farm, whether innate or otherwise. Secondly, he has to receive sound advice presented in a constructive, encouraging way with an understanding of, and an empathy with, farming. Thirdly, there has to be a cash incentive to compensate for any reduction in the agricultural production that might reasonably be incurred by implementing measures to cater better for farm wildlife. Whether such payments, as they are at present, should be *in addition* to the Basic Payment Scheme subsidies – the BPS – is discussed in other chapters.

Farmers such as Peter French have been working positively and constructively with advisers such as Cath Jeffs. The result is a reduction in farm productivity – though perhaps very little in profitability – but huge gains for wildlife. The cash incentive comes from Countryside Stewardship (CS), the agri-environment scheme in England. It has numerous land management options with applications scored on the basis of how much wildlife and other environmental improvement an applicant offers on his farm against priorities set for their part of the country.[v] In the Mid Tier of CS, applicants can select from no fewer than 134 options and capital items depending upon the type of farm they have and what they propose to do. There are no guarantees of acceptance, but if they are accepted, an agreement runs for five years.

With an emphasis on bird conservation, an obvious option for a farmer interested in entering the Mid Tier is SW1: establishing a four to six metre buffer strip of uncultivated, generally unsprayed, and unfertilised land around the outside edge of a crop, cutting the part nearest the crop annually but leaving the rest uncut except to reduce any woody growth. That option pays £353 per hectare of strip, although to guard against double funding a farmer using such a strip as an Ecological Focus Area (EFA) to get his greening payments (see Chapter 4) would receive a reduced CS payment of £79 per hectare. Implementing this measure would help to provide habitat for a range of birds (both rare and common), flowering plants, and invertebrates. Join up such buffer strips from field to field – hopefully across adjacent farms too – and mobile species would have opportunities to move to better feeding and breeding spots and in response to climate warming.

There are also Stewardship options for retaining winter stubbles after cereal and oilseed rape crops, particularly valuable for seed-eating birds in winter; for creating so-called 'Skylark plots' of un-sown soil within arable crops to provide nesting and feeding sites; for the management and gap-filling of existing hedges; and many more besides.

The scheme's Higher Tier is a ten year agreement that relates to farms that have environmentally significant sites, often designated as SSSIs; plus common land and woodlands where more complex management requires support from Natural England or the Forestry Commission. Here the options that would benefit birds and much other wildlife include: the management of upland moorland

(£43 per hectare); the creation of reedbeds (£323 per hectare); and the creation of wet grassland for breeding wading birds (£406 per hectare) such as Lapwing, Redshank, Curlew, and Common Snipe.

A LARK DESCENDING

Several farmland birds need as much Countryside Stewardship help as they can get! Take Skylarks, once the veritable sound of summer across the British lowland countryside, so plentiful that the summer sky seemed never to fall silent. In the 19th century only a few remote Scottish islands didn't sustain breeding pairs.[i] In spite of massive numbers being trapped by nets and killed for eating (plus lesser numbers captured as cage birds for their song) there was no noticeable decline through the early half of the 20th century until maybe the 1960s or more recently. Their breeding numbers declined by 58% between 1970 and 2010 although they are still very widely distributed across the country.[ii] Their numbers have fallen more precipitously, and are still falling, on farmland, due to the switch to autumn-sown cereals and much more intensive management of grasslands.

Autumn-sown cereals are taller and denser than spring-sown cereals throughout the breeding season because they have had longer to grow by the time the birds begin nesting. Research has found that fewer birds nest in them, and those that do are unable to raise as many broods as birds nesting in spring-sown crops.[vi] Many nesting attempts in autumn-sown crops are on or close to tramlines (the tractor tracks that are used to apply the many sprays to the crop), which makes the nests vulnerable to predators as they are so much more visible on the bare soil. Winter food is in short supply in the absence of stubbles – favourite feeding places replete with flowering plant seed, a bit of spilt grain, and plenty of invertebrates – while the increased use of insecticides and herbicides will kill off an important part of the food anyway.

In grassland habitats, farmland intensification has also been detrimental. Increased livestock numbers graze the grass too short for Skylarks and increase the risk of nests being trampled. The switch from hay to silage making has resulted in many nests being destroyed by the cutting machinery because the first cuts are taken so much earlier in the season.

Consequently, farming that includes spring-sown cereals, cereal stubble, hay cutting rather than silage making, and low-intensity pasture grazing is likely to increase Skylark numbers and will benefit many other farmland birds too. This is why Countryside Stewardship and the other agri-environment climate schemes around the UK include these measures in their selection of options. For instance

Any hope of comeback? Grey Partridge (courtesy of Bernard Stam).

Glastir, the Welsh agri-environment climate scheme, has 'retain winter stubbles' (Option 28 at £122 per hectare), and 'reversion of grassland to hay cutting' (Option 122 at £249 per hectare) which converts grazed pasture to annual hay cutting.[vii] Countryside Stewardship also has an option to create so-called 'Skylark plots' as feeding sites (option AB4), paid at the rate of £9 per plot;[v] each one requires an area of 16 square metres left unsown and scattered throughout a cereal crop so that sparse natural vegetation grows in them instead.

Research has found that such plots increase Skylark populations, their breeding density, the duration of breeding, and their breeding success compared with those breeding in crops without plots.[viii] Similar plots are encouraged by Countryside Stewardship in the parts of eastern and southern England where there are Stone Curlews to help provide them with patches of bare and thinly vegetated ground that these skulking ground birds need for breeding. Strange, almost reptilian-looking birds, with large eyes and frequently motionless posture and not at all related to Curlews (but given their name because of similar calls), they are still less common in Britain than the distribution of their preferred habitat of bare stony ground with very short vegetation might suggest.

Some studies, though not all, have also identified benefits to plants and invertebrates from 'Skylark plots'. There is also some evidence that Yellow Wagtails (yet another common farmland bird in decline) might find these plots helpful too, although cereal fields are most certainly not their first habitat of choice. They prefer damp pastures, and they have declined because those very places on most farms in the lowlands have been drained and often converted to arable crops, depriving these pretty birds of both breeding sites and of the insects they often feed on.

To show how effectively wildlife can be catered for on a productive lowland farm, the RSPB bought Hope Farm near Knapwell in Cambridgeshire in 2000. They are currently growing winter wheat, spring barley, winter beans, winter linseed, and spring millet on 160 hectares, with five hectares of permanent pasture grazed by horses and sheep. The farm receives the normal BPS and greening payments available to any farmer.

'We use two separate contractors to carry out the work here, one to carry out the cropping contract and another local farmer does the conservation management and small scale experiments on our behalf. My role is to liaise with them to ensure that the overall project delivers a high-quality farm which provides abundant wildlife while still producing good yielding and quality crops,' Ian Dillon, the RSPB's Hope Farm Manager says. 'We have a current Entry Level Stewardship agreement which terminates in September 2017. Within that agreement we deliver high-quality resources for wildlife in return for our payment which was £5,400 per annum before the 2013 CAP reform but has now dropped to £3,700. Our intention is to apply for entry to Countryside Stewardship when our current agreement comes to an end.

'Bidwells' – agri-business consultants – 'carried out a benchmarking project on our behalf last year comparing Hope Farm to a number of other similar sized farms, growing similar crops, within 50 miles of us. In the main we came out as average in this group of peer farms in terms of both production and profitability,' adds Mr Dillon.

In the 16 years since they bought it, Hope Farm's bird population has bucked the downward trend UK-wide in farmland birds' fortunes; their numbers have increased, not just a little, but substantially. In January 2016 they recorded 2,933 birds of 48 species, compared to 534 birds of 30 species in the same month when they took over the farm.[ix] Linnets, Yellowhammers, Reed Buntings, and Skylarks have at least tripled in number. Cutting hedgerows and ditches just once every three years and making insect- and seed-rich habitats have helped attract new species, such as Grey Partridge and Yellow Wagtail, to breed on the farm. Skylark breeding territories increased from 10 to 43 by 2015; Yellowhammer breeding territories from 14 to 35, and the numbers wintering from just two to a staggering peak count of 723. The increases are not only of birds; butterfly and bumblebee numbers have increased considerably, presumably due to there being more flowers in uncultivated margins at the edges of fields.

Similarly simple measures have been applied to other species such as Corn Bunting in eastern Scotland through encouraging mixed farming; creating annually sown, unharvested crop patches; and delaying the cutting of silage. 'Corncrake mown grassland' in the Scottish Government's agri-environment scheme is an option aimed at supporting the management of hay and silage fields for these ground-nesting birds. Corncrakes have declined catastrophically in Britain and are now confined to the western fringes of Scotland. The scheme includes reducing chick mortality by delaying mowing hay meadows and cutting them from the centre outwards, leaving them able to scurry into longer vegetation at field edges. It also includes providing areas of taller, flower-rich vegetation in spring for arriving birds (Corncrakes winter in southern Africa). The measures helped their numbers increase by 141% in the UK between 1993 and 2009, albeit almost exclusively still in the Western Isles.[v] The measure pays up to £460 per hectare per year depending on the agreed restrictions.

The Scottish Government's scheme also includes geographically targeted measures for other key birds. Two west coast islands – Islay and Colonsay, the Chough's only breeding locations in Scotland – are selected (based on RSPB advice) for mown grassland management without pesticides to provide the birds with good feeding habitat. Fields entered into an agreement by farmers on the islands must be grazed all year except in summer when one hay or silage crop can be cut. The farmer gets £224.75 per hectare per year. But while Chough (a species of crow with scarlet beak and legs) have increased across several coastal areas of the UK in recent decades, they have not re-established on the Scottish mainland. While part of this increase might be due to more sympathetic management by farmers of coastal grasslands, milder winters over the last couple of decades may equally be part of the explanation.

Glastir Advanced in Wales also has an option of 'Grassland management for feeding Chough' in which the height of the grassland sward has to be kept between three and five centimetres all year by livestock grazing without any pesticides, fertiliser, or vegetation burning. As with the equivalent Scottish option, the measure is available in mapped areas where these uncommon crows are known to breed and are dependent on short-cropped grassland to be able to probe the soil for the invertebrates they feed on. Unlike in Scotland, the farmer here receives £118 per hectare per year.

Targeted management of heathland in parts of Wales is reaping benefits for Black Grouse, as well as for upland heath habitat and the wide range of invertebrates and plants it supports more generally. The retention of stubble fields and growing wild bird seed crops has had a significant overall effect on many birds because these plants help fill the winter food gap. Birds that have benefited from these measures include Yellowhammer, Linnet, Reed Bunting, and Grey Partridge.

But are agri-environment schemes regarded as being successful? Research into their effectiveness in addressing declines in lowland farmland bird populations – especially where the causes of a population crash are well understood (such as for Cirl Buntings) – shows that remedial land management delivered by a scheme is successful when it has been tested, is well targeted, and is supported by good advice.[x] There is also considerable evidence that measures in the schemes targeted to improve the lot of birds such as Corncrakes, Stone Curlews, and Cirl

Food for winter finches. Maize stubbles retained until spring near Hereford.

Buntings, are resulting in increases in other species, ranging from carabid beetles and butterflies to Brown Hares and hoverflies.

The problem, though, lies in the scale at which they are operating and the fact that they are voluntary rather than being linked to the receipt of BPS (subsidy) payments. While there are local successes, there have yet to be measurable population recoveries UK-wide for farmland species including Skylark, Turtle Dove, Yellowhammer, Lapwing, and Grey Partridge.

In spite of the good take-up of agri-environment schemes across the UK, Turtle Doves – formerly widespread lowland farmland breeding migrants from West Africa – continue to decline, and have disappeared entirely from the west of Britain. The small population that remains breeds exclusively in the east and southeast of England; it isn't impossible that the species might become extinct here. It's not only the decline of 'weed' seeds amongst farm crops reducing their feeding ability in the breeding season, but other factors such as hunting when they are migrating and habitat loss in their winter quarters that are taking a toll of their numbers. British Trust for Ornithology (BTO) monitoring information shows a decline in numbers of an incredible 93% between 1994 and 2016.[xi]

In general, the more connected farmland habitats are, the more likely it is that the species that depend on them for feeding and breeding will prosper. This is simply because they can exploit different areas at different times if and when the quality of the habitat varies from place to place with changes in factors such as seasonality. The voluntary nature of agri-environment schemes often leaves gaps in that habitat continuity if only a proportion of farmers in a given area decide to join.

Some agri-environment measures are probably more effective than others, but there isn't long-term data to be absolutely certain which the best are. Short-term trends in bird populations are affected by issues that include changing weather patterns, so long-term data is essential to guide decisions about which measures are most effective and give value for money.

One study of Corn Buntings in North East Scotland for example found that targeted agri-environment measures and good advice converted widespread annual declines in non-scheme farms of 14.5% into increases on farms with agri-environment schemes of 5.6% annually.[xii] But the number of farmers entering their land for the schemes meant that only a quarter of the bunting population was being targeted, and the researchers calculated that they would need to target nearly three-quarters of it to halt the bird's decline on a national scale. There have been similar findings for other key lowland farmland birds such as Lapwings and Grey Partridge. Their numbers have been bolstered on farms taking up the schemes but the extent of these isn't enough – yet anyway – to even slow the overall decline nationwide simply because such populations are sinking fast on the larger area of intensively-managed farmland where there's no room left for them!

Some farmers, though, are working together as a group to implement conservation measures. Known as 'clusters', it's an idea developed in 2013 by the GWCT and is supported by Natural England. It's a common sense approach, seemingly gaining in popularity. By the end of 2016 there were 49 such clusters in England incorporating around 1,000 farmers and farm managers covering thousands of hectares of farmland. Many sign up to agri-environment schemes; others simply devise their own conservation plans having taken advice yet decide not to take any payments. The downside of this is that farmers might decide to encourage certain species of plants and animals but not others for which the conservation need might be stronger. Nevertheless, clusters are a significant step in making farming over much larger land areas more attractive to wildlife, something that previous 'pepper-potted' agri-environment schemes have failed to achieve.

There is little doubt that without (often simple) modifications to farming practices, almost always paid for out of agri-environment scheme money in addition to BPS and greening (farm subsidy) payments, some otherwise declining farmland birds would not have begun to recover their numbers. The Cirl Bunting on the South Devon coast is an excellent example. Other less rare but much declined farmland birds such as Tree Sparrows and Yellowhammers are also benefitting from the measures that are bringing back the buntings.

However, a recent study showed that agri-environment schemes have the potential to boost bird populations on farmland at a national scale although some scheme components had little effect.[xiii] As indicated earlier, a review of numerous studies concluded that such schemes 'can be spectacularly successful in reversing declines in farmland birds' provided the reasons for the decline are understood, that the scheme is well targeted, that farmers get sound advice, and the measures are delivered to a high proportion of its population.[x] It is the last point that is the most telling!

The 2016 State of Nature Report makes clear that:

> There is growing evidence that many farmland birds are benefitting from key environmental stewardship options, but others continue to decline and it is not yet clear whether stewardship can be delivered on a sufficiently large scale to achieve wildlife recovery nationwide. Certainly, at present, the hoped for widespread recovery of farmland wildlife is yet to be seen. The Farmland Bird Indicator (a set of 19 once common lowland farmland birds such as Lapwing and Tree Sparrow) shows a decline of 54% since 1970 (up to 2014), and although the rate has slowed in recent years, the decline continues.[xiv]

If agri-environment measures are going to be successful in reversing the declining fortunes of many farmland birds – and thereby many other species of plants and animals – by recreating farmland habitats and more appropriately managing those habitats that remain, they need to be adopted much more widely across large numbers of farms UK-wide, with habitats joined up farm to farm.

In Britain's uplands it will require more sensitive management over large tracts of moor, blanket bog, and mountain land; less vegetation burning; reduced livestock grazing overall; no grazing in particular areas; and conversion of some agriculturally-improved pasture back to more attractive wildlife habitat. In the lowlands, the job is often more fundamental and will include recreating lost habitats; reducing pesticide use; retaining and more carefully managing important features such as hedges, small areas of scrub, and rank vegetation; and managing the margins of much crop-growing land to provide breeding and feeding habitat for once-common farmland birds.

What is needed is more cash directed away from BPS subsidy payments in order to increase substantially the scale of take-up of agri-environment agreements. Continuing to run these schemes as voluntary measures might not provide sufficient take-up by enough farmers to redress countrywide the enormous losses of farmland birds that have occurred over the last few decades.

Endnotes

i   *The Historical Atlas of Breeding Birds in Britain and Ireland, 1875–1900*, by Simon Holloway. T. & A. D. Poyser, 1996.

ii   'The State of the UK's Birds 2016,' RSPB 2017.

iii 'Wild Bird Populations in the UK, 1970 to 2014,' Annual statistical release. Defra, 2015.

iv   *Bird Atlas, 2007–11: The breeding and wintering birds of Britain and Ireland,* by Dawn Balmer et al, British Trust for Ornithology, 2013.

v   'Countryside Stewardship Mid Tier Manual,' Natural England, 2016.

vi   'The effects of agricultural intensification on Skylarks: Evidence from monitoring studies in Great Britain,' D. E. Chamberlain and G. M. Siriwardena. *Environmental Reviews*, 2000, 8(2): 95- 113.

vii GLASTIR: Glastir Advanced 2017: Rules Booklet 2, Whole Farm Code and Management Options. Welsh Government, 2016.

viii www.conservationevidence.com/actions/540.

ix   www.rspb.org.uk/forprofessionals/farming/hopefarm.

x   'Agri-environment schemes and the future of farmland bird conservation,' by Jeremy Wilson and Richard Bradbury. *Wildlife Conservation on Farmland, volume 1*. Edited by David Macdonald and Ruth Feber. Oxford University Press, 2015.

xi   https://www.bto.org/news-events/press-releases/turtle-dove-population-tailspin.

xii 'Adaptive management and targeting of agri-environment schemes does benefit biodiversity: a case study of the corn bunting,' Perkins et al, 2011. *Journal of Applied Ecology*, 48: 514.

xiii 'Landscape-scale responses of birds to agri-environment management: a test of the English Environmental Stewardship scheme,' D.J.Baker et al, 2012. *Journal of Applied Ecology*, 49:871–882.

xiv 'State of Nature 2016,' D.B.Hayhow et al. The State of Nature Partnership, 2016.

CHAPTER 8

RETAINING THE FOUNDATIONS

The nation that destroys its soil destroys itself.

FRANKLIN D. ROOSEVELT. FROM A **LETTER TO ALL** US STATE GOVERNORS ON A **UNIFORM SOIL CONSERVATION LAW,** *26 -FEBRUARY 1937*

There is one wildlife habitat on farmland that is far richer in species than anything else; richer than hay meadows, hedgerows, broadleaved woodland, or anywhere else that is replete with a broad range of plants, invertebrates, small mammals, and birds. It underpins everything a farmer seeks to do with his land and which he neglects at his peril. It is, of course, soil.

Most of us think of soil as the medium in which plants grow; a black or dark brown material typically consisting of a mixture of ground up rock particles, organic matter, minerals, air, and water. It's the organic matter and minerals, together with added fertilisers and organic manures, that provide the principle nutrients any farm crop needs in order to grow. But soil is much more than a medium in which to grow plants.

Soil is the most abundant ecosystem on Earth; estimates suggest that one gram of it can contain perhaps a billion organisms including five million bacterial cells, 10,000 protozoa, 200 metres of fungal hyphae, and around 100 nematodes.[i] There are so many species in the soil; many have not even been identified. It's very likely that soil harbours the most diverse populations of bacteria of any environment on earth. Soil organisms range in size from tiny one-celled bacteria, algae, fungi, and protozoa, to more complex organisms like worms, insects, small vertebrates, and plants. Along with earthworms, arthropods, mites, springtails, and beetles, this massive array of organisms play an important role in maintaining soil health.

Their numbers, though, vary enormously depending on soil type, location, and depth.

Underneath arable crops there might be five tons of living organisms per hectare; under grassland it might be 20 tons per hectare! Yet the 2016 State of Nature report doesn't even mention soil wildlife!

LITTLE ROTTERS

The first major difference between the above-ground food web and the soil food web is that the soil food web has a different source of energy. Above ground, the energy source is the sun; below ground, the energy source is decaying organic matter. Organic matter – crop residues for example – is manna from heaven for the decomposers that live in soil, bacteria and fungi especially, and when bacteria and fungi feed on such organic materials they release nutrients back into the soil that aid plant growth.

These nutrients help to make humus, the dark, nutrient-rich part of the soil. Therefore, decomposition not only provides food for decomposers, it also helps to create more soil. Humus is 'new soil', and so it is extremely important. It provides nutrients for the plants, shrubs, and trees that grow on its surface. The new soil also helps support many of the animals that live underground, like worms, by providing these organisms with a place to live and a nutrient-rich food supply. Decomposition also helps to conserve water in the soil and limit soil loss.

Without these little rotters, the piles of crop stems and leaves ploughed back into the soil after harvest would simply pile up so high that the land would be impossible to farm. The rotters are the original recyclers and we damage or destroy them at our peril

Bacteria and fungi in the soil are also food for larger soil organisms: mites, springtails (nearly 400 species of them in Britain), eelworms, and lots of others. In turn, these little creatures are food for birds such as Starlings, Curlews, and several more that probe into the very top layers of soil.

The larger soil animals like earthworms mix soils as they move around, forming burrows and pores as they go, thereby allowing moisture and gases to move about. In the same way, plant roots open channels in soils. These pore spaces are abundant in well-structured soils, comprising over half the soil volume. Plants with deep taproots can penetrate many metres through different layers to bring up nutrients from deep down to near the surface. Plants with fibrous roots that spread out near the soil surface have roots that are easily decomposed, adding more organic matter. Many microorganisms, including fungi and bacteria, cause chemical exchanges between roots and soil and act as a reserve of nutrients. Mycorrhizal fungi form mutually beneficial associations with plant roots; the fungi gain sugars and other carbon compounds produced from the plant's photosynthesis and, in turn, provide the plant with minerals, nutrients, and water garnered from a much larger area of soil as their delicate hyphae spread out.

Earthworms ingest soil particles and organic residues, enhancing the availability of plant nutrients in the material that passes through their bodies. They aerate and stir the soil, thereby assuring ready infiltration of water. Crops growing in soil remove nutrients from it as they do so but, like natural vegetation cover which holds the soil in place, they also help to prevent its erosion by heavy rain that might otherwise result in surface runoff. Plants shade soils, keeping them cooler and slowing down the evaporation of soil moisture; conversely, though, by way of transpiration plants can cause soils to lose moisture.

WHAT'S A GOOD SOIL?

It's one that:

- Drains well and warms up quickly in the spring
- Does not crust over after planting
- Soaks up heavy rains with little runoff
- Stores moisture for drought periods
- Has few clods and no hardpan on or near its surface
- Resists erosion and nutrient loss
- Supports high populations of soil organisms
- Has that rich, earthy smell
- Produces healthy, high quality crops and grass
- Is easy to work in a range of conditions.

(Source: LEAF (Linking and Environment Farming) 'Simply Sustainable Soils'.)

But are we treating farm soils correctly? Are we conserving them in order to sustain their natural wealth so that they will continue to be capable of growing the crops we need to feed ourselves? Are we protecting the incredible variety of life forms that populate them, or allowing them to be depleted like many above-ground plants and animals on farmland?

Instant recognition: Lumbricus terrestris, the Common European Earthworm and the best known soil creature (courtesy of Michael Linnenbach).

The UN's International Year of Soils was in 2015, and aimed to highlight the importance of healthy soil and draw attention to the threats facing this vital resource. Speaking in April 2015 at the third Global Soil Week in Berlin, Moujahed Achouri, Director of the UN's Food and Agriculture Organisation's Land and Water Division, called the rate of soil degradation around the world 'alarming'. Globally, a third of agricultural soil is classified as degraded. A 2013 report by the EU's Joint Research Centre revealed that soil biodiversity is under threat across 56% of the EU's land area.[ii] Yet in 2014 the European Commission was forced to withdraw its proposals for a Soils Directive, which would have put legal obligations on farmers to conserve soils and establish land management practices to reduce its erosion and increase its organic matter content. The proposal sounded like common sense but the UK, Germany, France, the Netherlands, and Austria were firmly against it. They blocked it from going ahead, reflecting the views of many farmers and the industry, including the UK's National Farmers' Union (NFU), who were strongly opposed to additional legislative controls over soil protection. Why? They saw it as a burden on farmers.

Dr Maria Tsiafouli and her colleagues have examined the biodiversity in soil under grasslands and intensive crop rotations on farmland in four European countries (including the UK) and found that intensive farming reduces soil biodiversity by as much as 50%.[iii] Soil-dwelling earthworms, mites, and springtails in particular declined, all of which play direct or indirect roles in the decomposition of plant material in the soil. The authors go on to conclude that these changes in soil biodiversity due to farmland intensification may eventually threaten the functioning of farmed soils. 'The loss makes agricultural land less healthy and more vulnerable to pests and diseases,' says Dr Tsiafouli. That's something the farming industry needs to take heed of.

Other studies have found that the content of organic matter in intensively-farmed soils has declined, which has substantial implications for soil's ability to hold water and nutrients, for soil strength and its erosion, for carbon dioxide release to the atmosphere, and the pollution of waterways.[iv] In the first study of its kind, Dr Jill Edmondson and colleagues found that soils under Britain's garden allotments are significantly healthier than soils that are intensively farmed.[v] Allotment soil had a third more organic carbon, a quarter more nitrogen, and was significantly less compacted. Three-quarters of surveyed allotment plot holders added manure to their soils, 95% composted waste plant material on-site, and many added organic-based fertilisers and commercial composts. They were recycling nutrients and carbon back to their soil more effectively. Commenting on the figures, Dr Edmondson praised the management of soil by allotment holders while highlighting how damaging modern agricultural practices are.

Organic matter is the lifeblood of a healthy soil, something that most gardeners take very seriously. It gets into the soil through the decomposition of plants on the soil surface (the stems and leaves after a crop has been harvested on farmland), from root exudates, and from living and dead soil organisms. It can be added to soil, for instance, by manuring fields and allowing animals such as earthworms to incorporate it into the soil. Once in there, an army of tiny rotters decompose it to provide humus, the natural organic matter in a healthy soil. Roughly ten kilos of crop vegetable matter on the surface of the soil decomposes to about one kilo of humus within it. As it decomposes, nutrients are released back into the soil and become available for the next farm crop to grow.

However, humus does a lot more than that: it helps conserve water in the soil, aids in reducing compaction and surface crusting, and improves soil structure, thereby increasing water infiltration into the soil. The problem is that soil organic matter is too often ignored and neglected. To retain the necessary amounts, frequent applications of manure and compost are needed, as continuously harvesting crops actually depletes the soil humus unless large amounts of un-harvested stems and leaves are ploughed back into the soil.

For farmers, the only provision relating to maintaining the organic content of soil that's mandatory if they are claiming BPS subsidies is that they must not burn cereal straw or cereal stubble or crop residues from oilseed rape, field beans, or peas. This ensures that such crop residues are ploughed back into the soil to help bolster its organic content.

Intensive cultivation removes nutrients from the soil year after year; most farmers compensate for this by adding fertilisers to try to maintain soil fertility. For the last 50 years or so, soil fertility has been regarded as merely the result of a simple chemical test and a calculation about how much nitrogen (N), phosphate (P), and potash (K) to add as fertiliser. Mostly it's added in the form of synthetic inorganic fertilisers which also add some other nutrients in trace quantities. Organic farmers, though, can use only natural fertilisers: animal manures, composted plant matter, urea, and a few others.

To be able to hold and recycle nutrients and water, soil needs to be biologically active and humus-rich, a mantra for organic farmers. As humus levels increase, the structure of the soil becomes lighter and more friable so it has a greater ability to hold nutrients and water. High humus soils are capable of absorbing much more rain, thereby reducing run off and erosion. Excess water percolates through the soil and has its nutrients removed before entering ground water and getting into streams and rivers where it would otherwise pollute them and reduce their diversity of plants and animals. Retaining more water within the soil for future crop plant use helps to drought-proof them in periods of prolonged dry weather.

GETTING AT THE ROOT OF IT

Mycorrhizal fungi form mutually beneficial associations with plant roots; their incredibly fine fungal hyphae get inside the plant root cells or remain closely adhered on the outside of them. The fungi gain sugars and other carbon compounds produced by the photosynthesising plant. Meanwhile, the fungi provide the plant with minerals and nutrients gained more readily from the soil, particularly phosphates, because the fungal hyphae are so much finer than the plant's roots and have a much larger surface area. Mycorrhizal fungi also help supply water which is transported along the surface of the fungal hyphae to the plant roots.

Spraying crops prone to fungal diseases (especially in the west of the UK where it is wetter) risks killing off beneficial fungi including mycorrhizal species in the soil. A review by the European Commission examining what pesticides can be used on farms could result in a ban or restrictions in the use of azole fungicides due to fears they might act as endocrine disruptors, meaning that they could interfere with human hormone levels. Azole fungicides have been the backbone of crop fungicide treatment since the mid-1970s due to their broad spectrum of disease control and their systemic action throughout treated fungi. But concern has been growing that plant fungal pathogens are building resistance to fungicides, as they are to other pesticides, and experts such as Dr Neil Paveley, Head of Crop Protection at ADAS (the UK's largest independent agricultural and environmental consultancy) are advising farmers to reduce the number of fungicides that they spray in order to slow the development of resistance to them.

In general, lowland farm soils don't get much of a rest because crops aren't rotated as frequently as they once were, for example, from arable to pasture (supporting manure-producing livestock) to root crops and back to arable over a four year cycle. Consequently, the organic matter content of many farmed soils has declined, and with it the array of soil-living organisms. Organic farming encourages adding composted material, livestock manure, and green manures (plants that naturally supply nutrients) to soil. Green manure is created by leaving uprooted or sown crop parts to wither on a field so that they serve as mulch. The plants used for green manure are often cover crops grown primarily for this purpose. Typically, they are ploughed in and incorporated into the soil while green or shortly after they finish flowering. Clovers and vetches are good examples; as they are legumes they add both nitrogen and bulky organic matter to the soil when they are ploughed in, a double benefit.

For farmers with large bank loans wanting to clear any debts as soon as they can, short-term exploitation of soils to produce as much crop as possible sometimes triumphs over any consideration of the long-term sustainability of soil to continue to produce good crops. Some farmers are growing crops that are simply incompatible with protecting the soil. Others maybe don't have enough knowledge about soil erosion and the organic and nutrient content of soils. A few, especially contractors working on other people's land, don't seem to care.

A study in South West England between 2002 and 2011 identified soil structural degradation to be widespread, with 38% of the 3,243 surveyed sites having soil degraded enough to cause surface-water runoff.[vi] Soil under arable crops often had high or severe levels of structural degradation, and three-quarters of maize crops had damaged soil causing runoff of soil and water. This is mainly because maize harvesting has to be done late in the season when very wet conditions are most likely. One way of reducing this runoff is to sow a grass crop with the maize so that when the maize is harvested, the soil is still covered. Farmers having soil erosion taking place over any single area greater than one hectare can fail a so-called cross-compliance inspection and have their BPS subsidy payment reduced, albeit by very little.

Maize cultivation has expanded enormously in the UK over the last few decades, much of it for cattle feed rather than human consumption. A considerable amount is also used for fuelling anaerobic digesters (AD) to produce biogas as a source of fuel on farms, an innovation the EU has encouraged but which was intended to use up polluting manure slurries and crop wastes, and not as an excuse to grow more crops specially to fuel digesters. At present the UK has just 18 AD plants running using only manure slurry, though many more are planned.

Between 2008 and 2015, the area of crops grown for feeding AD units – mostly maize but rye and sugar beet too – expanded over threefold to 93,000 hectares. Defra has been subsidising digesters using maize, although in December 2016 it announced significant reductions in these incentives as part of a broader review.[vii] To what extent this will decrease the amount of maize grown in England is impossible to predict, but any reduction in such a potentially environmentally damaging crop is to be welcomed.

Soil erosion by water and wind can be a big problem for some farms, as it removes fertile topsoil, clogs drainage systems, and pollutes watercourses. Rain percolates into soils whose structure is intact but runs off fields where the soil structure has broken down, taking some of the soil – and any pesticides and fertilisers it contains – into rivers and other watercourses where it deposits silt on stony river beds (damaging fish spawning habitat) and pollutes the water itself, thereby depleting its natural communities of plants and animals. Poaching from livestock gathering around feeding spots and at access points to fields can increase the risk of soil erosion and contribute to soil loss. Pigs and poultry kept outdoors can cause trampling and compaction of topsoil, erosion, and runoff on slopes too.

In some parts of England, peaty soils – such as those in the Fens in the east, which were drained centuries ago – are farmed intensively to grow cereals and

Maize: a crop with a poor record for soil and water runoff (courtesy of Derek Ramsey).

vegetables. However, the now-drier peat oxidises because it's exposed to the air rather than being kept waterlogged and stable as it would be naturally. In consequence, the soil level in the Fens is falling by over two centimetres a year and the region will eventually become far less agriculturally productive.[viii] Over the last couple of centuries, fenland has shrunk several metres in height.

Defra estimates that soil degradation costs England and Wales an estimated £0.9 billion to £1.4 billion per year:

> While rates of soil erosion in England are not excessively high it's estimated to directly affect around 17 per cent of land in England and Wales but also create off-farm costs to the environment. Around 3.9 million hectares of our soils are also at risk of compaction which could lead to a total crop yield reduction of around £163 million per year in England and Wales.[ix]

Catchment Sensitive Farming (CSF) is a project run by Natural England in partnership with the Environment Agency and Defra, and has its origins in the EU Water Framework Directive which became law in 2000. The Directive promotes a new approach to water management through river basin planning in order to prevent deterioration in water quality; improve and protect inland and coastal waters and groundwater; lead to better and more sustainable use of water;

restore wildlife habitats; and help reduce the effects of floods and droughts. Soil, fertilisers, and pesticides from farmland contribute to water pollution and CSF is aimed at raising awareness of water pollution from agriculture by giving free training and advice to farmers in selected areas in England in order to improve the environmental performance of farms.

Ideally, it shouldn't need legislation or regulations to force farmers to take great care of their soil. It should be second nature. Gardeners growing vegetables and fruits, and allotment holders doing the same, hardly need to be told to safeguard their soils, add plenty of organic matter, and ensure that the very substance from which their produce develops doesn't wash away! Yet the standard of farm soil conservation in the UK's lowlands, as in much of the EU, is not what it should be. Until 2015, the only requirement a UK farmer had to meet in order to claim his BPS subsidy was to produce a 'soil protection review': a very brief field by field note of how soil is being protected on his farm, a paper or online exercise rarely checked by inspectors. The Rivers Trust – the charitable organisation promoting 'sustainable, holistic and integrated catchment management and sound environmental practices' – has previously called it an unenforceable exercise.[x]

From 2015, the new rules for farmers claiming BPS payments (in other words, virtually all of them) require no paperwork, but they have to maintain a 'minimum soil cover'; limit soil and bankside erosion (by preventing livestock from overgrazing for example); curb soil compaction; and sustain good levels of soil organic matter. 'After harvesting a crop, leaving the crop stubble and not ploughing it in is sufficient soil cover,' comments Simon Draper of the Farming Advice Service (FAS). The FAS is an organisation funded by Defra to help farmers understand and meet the requirements of cross compliance, greening, and the European Directives on both water protection and sustainable pesticide use. 'If he ploughs in the stubble, a farmer has to sow a cover crop (such as vetches or clover which can be eventually ploughed back into the soil) within 14 days or as soon as practicable. The Rural Payments Agency in England – and their equivalents in the other countries – can inspect a farm to check compliance; they check 1% of claimants annually and can impose fines.' That means a check on an individual farm on average once a century! Despite this, Mr Draper does point out that any farmer seriously contravening the requirements is likely to be noticed in other ways, by the Environment Agency for polluting a river perhaps, so inspections would then be more frequent and fines would escalate rapidly.

The requirement to retain soil cover, designed to reduce both water and wind erosion of soil is, however, providing an unexpected boost for some birds. Simon Draper estimates that there has been a substantial move by lowland farmers back to spring sowing of cereals rather than autumn sowing. The stubble qualifies as a cover crop and provides seed-eating birds with spilt grain and weed seeds through the winter. That's a huge benefit to many small birds, both seed-eaters picking up 'weed' seeds in the stubble and any spilt grain, plus insect-eating species after small invertebrates on and near the soil surface.

Soil compaction occurs when soil particles are pressed together; this decreases the natural pore space between them, which thereby reduces water infiltration from rain and drainage of water from the soil. It's often very visible after rain in arable fields where water fails to drain away in the tramlines caused by tractor and equipment wheels compressing the soil. The exchange of gases slows down in compacted soils, cutting the aeration that is essential for crop root growth. It also means that crop roots must exert greater force to penetrate the compacted layer. As farm tractors and field equipment have become larger and heavier over the last few decades, concern about compaction has grown. A field tractor today can weigh up to five times its 1940s counterpart!

Not all soil compaction is bad news. Slightly-compacted soil can speed up the rate of seed germination because it promotes good contact between the seed and soil; moderate compaction may reduce water loss from the soil due to evaporation and, therefore, prevent the soil around the growing seed from drying out. But excessive soil compaction impedes root growth and thus limits the amount of soil explored by crop roots. This in turn can lower the plant's ability to take up nutrients and water. In dry years, soil compaction can lead to stunted, drought-stressed plants due to less root growth; in wet years decreased soil aeration can cause a loss of nitrate from soils. In both cases, crop yields fall.

Measures such as not cultivating wet soil; reducing the number of times the ground is compressed with equipment; regularly checking fields for signs of compaction; and loosening the soil after harvest, especially along tramlines, are all practical steps that can help reduce soil compaction and soil and water runoff.

Most farmers do manage their land sensibly to reduce erosion, compaction, and other soil degrading problems. For the minority who do not, regulation backed up by meaningful fines is a must. But farmers and their representatives are set against yet more regulation. Lobbying on behalf of the UK's farmers – led by the NFU – has influenced successive governments over the last few decades to limit legislation and give them a freer hand to get on with their job. It's an easy theme often taken up by government ministers, most recently by Andrea Leadsom, then Secretary of State for Environment, Food and Rural affairs. When she spoke at the Oxford Farming Conference in January 2017, she proposed 'scrapping the rules that hold us back, and focusing instead on what works best for the UK' post CAP.[xi] With other EU countries being the major market for UK farm produce, a continuation of this essential trade when we have left the Union is very likely to require the UK's farmers to adhere to all EU farming standards anyway. The trick might be to make their implementation less bureaucratic, although that might be a lot easier in theory than it is in practice.

Doubtless the NFU and the farming unions in the other UK countries (the Ulster Farmers Union, the National Farmers Union Scotland and the Farmers Union of Wales), will take the opportunity provided by the UK leaving the CAP to argue for a more 'trust the farmer' agricultural policy. The Farming Regulation Task Force's report to the Government on ways of reducing regulatory burdens on farmers and food processors, published in 2011, recommended more than 200 ways of reducing unnecessary 'red tape'.[xii] Promoted as an 'independent' review, it was chaired by a former Director-General of the NFU and all except one of its members were farmers or were intimately involved in the food processing industries; not a composition likely to provide much alternative thought.

One way to reduce soil erosion is to never plough the land, instead directly drilling the next crop's seed into the top layer of the soil and leaving the previous crop stubble on the surface. Any soil mixing is left to crop roots and the natural soil fauna. Eliminating tillage reduces soil compaction and erosion because heavy equipment is driven less often over the soil, and it doesn't expose areas of soil directly to wind and heavy rain. Direct drilling also reduces fuel costs for obvious reasons and it leaves more crop residue on the soil surface which further limits erosion. It also has sizeable advantages for farm wildlife above the soil. Leaving crop stubbles or sowing cover crops provides more habitat for invertebrates and for feeding flocks of birds, including finches, that devour any grain spill from the harvested crop. The soil biota benefit too; unploughed or very lightly ploughed soils have more organic matter, more aeration, and are less likely to experience the extremes of drying out or waterlogging.

Anthony Pope, an experienced agricultural consultant who advises LEAF (Linking Environment and Farming) is an advocate of no-till farming. 'I believe zero tillage has huge benefits and will significantly replace ploughing in the future, but UK Government policy, research, academics, machinery manufacturers, and lead farmers will all have to work together to make it happen. It does mean that farmers have to be more knowledgeable with a greater understanding of soils with more attention to overall land and crop management detail. We've moved from ploughing to minimum tillage over the last 20 years but what that's done is to raise the level of the compacted layer in the soil (because heavy machinery is still used) and left a ten centimetre layer of dusty soil on the top which can't support crops in dry spells. Minimum tillage hasn't brought any environmental benefit either. Crop yields might dip for the first few years after conversion to no till but they are then on a par with conventional systems. The much lower input costs provide significantly better profitability,' he says.

John Cherry, who farms over 800 hectares near Stevenage in Hertfordshire, is firmly committed to no till. 'We're mostly on chalky boulder clay which used to be quite tricky when we cultivated, but it's easier with no till. We have a bit of chalk too which is very much easier and is getting darker now the soil organic matter is building up. There's no doubt that lighter land is much easier for no till and benefits more from absorbing rain to help the crops grow. We are now in our sixth year of continuous no till and are finding it gets better and better the longer we persist, but some newcomers are baulking at the problems they are encountering (for instance plagues of slugs). We tend to advise farmers to drill a fortnight earlier in the autumn than they would if they were cultivating but a fortnight later in the spring because the soil needs to warm up. We had problems with slugs to start with, but now we don't use any insecticides and we have built up slug predator numbers so we haven't needed to use any slug pellets this year,' he says.

No till even made it on to *The Archers* on Radio 4 during 2016, a clear sign that it's starting to get into the farming mainstream! It was Adam Macy's latest idea, albeit making Brian Aldridge feel a touch nervous. First it was herbal leys and 'mob grazing' with cattle, both designed to raise the long-term fertility of the soil, followed by Adam promoting no-till cereal growing.

THE END OF THE PLOUGH?

Anthony Pope lists the many advantages of no till farming:

- Improves soil organic matter, soil porosity, and soil organisms

- Lessens the effects of soil compaction

- Increases the availability of plant nutrients and uses less fertiliser

- Controls weeds and pests with fewer agrochemicals, provided that intelligent targeted techniques are used

- Enhances soil water absorption and groundwater recharge which decreases runoff and flooding

- Minimises labour and tractor use

- Reduces greenhouse-gas emissions from the soil and limits farm fuel use

- Lowers costs and increases profits

Defra doesn't collect any statistics on no-till farming but Tony Reynolds who farms in South Lincolnshire (and has been no till for 16 years) and represents the Association for Conservation Agriculture who promote it, estimates that around 322,000 hectares of UK arable land (over 7% of the total) is no till. That's risen from 150,000 hectares in 2011 and continues to increase. Most oilseed rape is sown without ploughing. In the US though, he quotes figures of 25% of arable land farmed with no tillage and of 60% in South America. Defra quotes figures from trials by Cranfield University showing very similar crop yields with no-till farming compared to conventional ploughing but a reduction in costs from £85–£120 per hectare for ploughing to £30–£45 per hectare for direct drilling, plus a substantial reduction in labour time.[xiii]

At present glyphosate is usually used to kill off the remains of the last crop and any weeds before the next crop is drilled into the soil. While the herbicide decomposes in soil, though at extremely varying rates, there have been recent concerns that traces of it in foodstuffs might present a carcinogenic risk to humans. Its use is banned or restricted in some countries and its toxicity has recently been assessed by the European Chemical Agency (Echa) after EU countries failed to agree to re-authorise its use. In March 2017, Echa concluded controversially that 'the available scientific evidence did not meet the criteria to classify glyphosate as a carcinogen, as a mutagen or as toxic for reproduction' though the European Commission still has to decide whether to accept Echa's conclusion.[xiv] Effective only on actively growing plants, its use has burgeoned in the last few decades. Glyphosate kills all growing plants, many species of soil bacteria and fungi, and some earthworm species, but its toxicity to most mammals is generally regarded as low. Whilst it is less persistent in water than soil, it has been shown to be toxic to some amphibians although seemingly not to aquatic invertebrates.

Agriculturally, there are substantial concerns that many plants have developed a resistance to glyphosate and that others will do so, increasing the need for alternative action. As John Cherry comments: 'We are all trying to find a way to farm without chemicals, but at the moment glyphosate is easily the best way of terminating cover crops or grass leys in no-till farming. There is a lot of indignation about glyphosate, but it is much less harmful than many pesticides we use and to my mind much less harmful than cultivating the soil. Pleasingly, there's a lot of research going on in the US about destruction of crops without chemicals – organic no-till effectively – but no-one has successfully transferred it to the UK as far as I know. Maybe we'll be farming with more tolerance of weeds as we learn how to and as soil fertility builds up so crops are healthier; this is certainly what we are working on.'

Jake Freestone, Farm Manager on the Overbury Estate in Gloucestershire is an advocate of no till and direct drilling of crops on the land he manages. He considers it a way of increasing both microorganism and earthworm activity in the soil while retaining its structure, thereby reducing the problems associated with leaching and compacting whilst better retaining soil moisture and eliminating soil erosion. It's a win/win in his books.

What has become very clear is that we can't afford to play fast and loose with our soils. They are the foundation upon which our crops and our livelihoods depend. It is usually forgotten in the welter of concern for farmland birds, bees, and other wildlife that our soil is the habitat supporting the richest array of living creatures on earth. Just as we should take better care of above-ground habitats – woodlands, hay meadows, wetlands, and the rest – whether underlying farmland or elsewhere, we need to look after our soils and take greater care of them than we do on much of our farmland at present.

Some existing measures such as planting soil cover crops to provide ground cover after main crops have been harvested will undoubtedly help minimise soil runoff and wind erosion of bare soils. But the UK's four agriculture departments and the farming industry itself need to consider more actively recommending modified land management such as no till to replace ploughing more widely in order to improve soil organic content and its biota while further reducing compaction and runoff. Better slurry application techniques and practices are also essential. Leaving such an important issue to voluntary application might not be sufficient.

Endnotes

i   'Farm Soils Plan: Protecting Soils and Income in Scotland,' Scottish Government, 2005.

ii   'FAO: Soil degradation reaching critical point,' *Farming Online*, 23 April 2015.

iii   'Intensive agriculture reduces soil biodiversity across Europe,' M. A. Tsiafouli et al, 2014. *Global Change Biology* 21, 973–985.

iv   'Soil organic matter stratification ratio as an indicator of soil quality,' A. J. Franzluebbers. *Soil & Tillage Research* 66 (2002) 95–106.

v   'Urban cultivation in allotments maintains soil qualities adversely affected by conventional agriculture,' Jill L. Edmondson et al, 2014. *Journal of Applied Ecology*, 51(4): 880–889.

vi   'Soil structural degradation in SW England and its impact on surface-water runoff generation,' R. C. Palmer and R. P. Smith, 2013. *Soil Use and Management*, 29: 567–575.

vii   'The Renewable Heat Incentive: A Reformed Scheme. Statement of policy and Government response to consultation,' Department for Business, Energy and Industrial Strategy, 2016.

viii   'An estimate of peat reserves and loss in the East Anglian Fens,' Cranfield University, 2009.

ix   'Agriculture in the UK 2015,' Defra and devolved governments, 2016.

x   'The farming lobby has wrecked efforts to defend our soil,' *The Guardian*, 5 June 2014.

xi   https://www.gov.uk/government/speeches/environment-secretary-sets-out-ambi tion-for-food-and-farming-industry.

xii   https://www.gov.uk/government/.../independent-farming-regulation-task-force-report.

xiii   'Crop Establishment to Protect Soils: Direct Drilling. Best Practice Information Sheet,' I S 2.3.2. Defra, 2011.

xiv   https://echa.europa.eu/-/glyphosate-not-classified-as-a-carcinogen-by-echa.

CHAPTER 9

THE CALL OF THE HILLS

Nature's heart beats strong amid the hills.

RICHARD MONCKTON MILNES (1809–1845), ENGLISH POET,
PATRON OF LITERATURE, AND POLITICIAN

Whilst there were occasional glimpses of sunshine on this late-April day down in Pateley Bridge in the Yorkshire Dales, it's certainly not shining up here! On the five kilometre track to Low Riggs farm, rather inaccurately named for an isolated, stone-built farmstead nearly a thousand feet up on the wet, peat-based moors above the Dale, the hailstones are coming down almost horizontally, low clouds sweep around us, and it's bitterly cold. I'm driving with Tara Challoner, Farming and Wildlife Officer with the Yorkshire Wildlife Trust (YWT). The YWT is part of the Nidderdale Partnership set up to work with farmers and others to conserve and enhance the area's rich historic, environmental, and cultural heritage.

Given the cloud and hail, I can see very little in these conditions, but in spite of the howling wind, I can just hear the occasional haunting call of a Curlew echoing across this open country. As the wintry weather clears for a few minutes, as well as spotting a couple of Oystercatchers flying low over boggy ground and a handful of Lapwings feeding in some pastures nearby, Ms Challoner points out the distant figure of Andrew Hattan, sensibly clad in full waterproofs, approaching our jeep on his quad bike.

Andrew and Sally Hattan took over the tenancy of Low Riggs in 2007; their goal was to create a farm that cared for the land and was financially sustainable in the long run. In the warm farmhouse kitchen out of the squally hail and biting wind, I'm discussing with Mr Hattan and Ms Challoner how farming at Low Riggs is managed to provide the habitat that upland breeding waders and other typical wildlife up here need.

BLANKETS OF PEAT

Blanket bog clothes around 1.5 million hectares of the British uplands, most of it in Scotland but also in the uplands of Central and North Wales and stretching along the Pennine spine of England. Wild, windswept, and undulating areas of spiky rushes, white-flowering cotton grasses, Cross-leaved Heath, and Ling, with carpets of squidgy sphagnum and other mosses all growing on waterlogged peaty soils, blanket bogs are a familiar feature of our dramatic upland landscapes. They have developed on a layer of deep peat, the part-decomposed remains of the very vegetation itself.

They are the most extensive type of wetland we have; they are fed entirely by rainfall and snowmelt rather than by groundwater. They are the giant sponges of the uplands, holding vast volumes of water that gets released slowly into the streams that drain into our larger rivers. Blanket bogs play a crucial role in water management: they provide up to 70% of England's drinking water.[i] Degraded bogs release large quantities of dissolved peat that water companies spend significant amounts of money removing from drinking water supplies and which can damage fragile pebble riverbeds; habitats for invertebrates and fish breeding.

Tucked away in their deep vegetation are more unusual plants and animals adapted to this specialised habitat, including Bog Rosemary, Cranberry, and insect-eating sundews. Invertebrates such as spiders, crane flies, and much less common species such as the Bog Hoverfly can be abundant. On drier slopes, large areas of moorland dominated by heathers and Bilberry take over. The whole mix provides breeding places for birds such as Golden Plover, Oystercatcher, Dunlin, Hen Harriers, Curlew, and Greenshank, alongside a wide range of aquatic and terrestrial invertebrates including around 40% of the UK's Large Heath butterfly population.

Much of this blanket bog and moorland has been planted with conifers, burned too regularly and drained by digging channels into the peat (so-called 'grips') or overgrazed with too many sheep. The plant diversity is then reduced, resulting – as it has in much of Central Wales – in large areas dominated by a dense cover of Purple Moor Grass, few other plants, and a paucity of invertebrates and breeding birds.

Blanket bog is slow growing; overgrazing it can release carbon dioxide through the loss of vegetative cover and lead to its surface layers drying out and eroding. Light grazing is necessary, though, because its natural climax vegetation – except in the wettest spots – is scrub, and without grazing or some other form of vegetation control the whole character of this landscape would change.

Taking advice. Andrew Hattan at Low Riggs Farm, Nidderdale with Tara Challoner.

'We're a 190 hectare farm but only about 20 hectares of it can be cut for silage and a few hectares for hay,' says Mr Hattan. 'We have 350 Dalesbred and Swaledale sheep plus another hundred on the top moorland where we have grazing rights. We don't use any artificial fertiliser; just some chicken manure on a few fields. Most of our land is heather moor or blanket bog and rough upland grazing, all of it pretty wet and much of it permanently waterlogged. We have some Galloway cattle but they're no good on the peaty vegetation with such high rainfall and it's impossible to finish cattle up here; they have to be sold off and put on better land to fatten them up for market,' he says.

The Hattans are in the second year of a Higher Tier Countryside Stewardship agreement with Natural England which runs until 2024, and they have been making numerous changes on the farm to improve both its viability and its wildlife value. Their first task was to rebuild the farm landscape and infrastructure: 2,000 metres of dry stone walls needed repair, and several hectares of ghyll (deep ravine) woodland needed restoring. Moreover, the upland hay meadows needed large areas of invading rushes killed off and their vegetation diversified by harrowing the pasture surface to create some bare ground, scattering flower seed harvested from similar meadows elsewhere in the Dale, and letting grazing cattle trample the seed into the soil. It's worked well, and the regenerated wet pastures have a richer flora as a result: frothy cream flower-heads of Meadowsweet, Wood Cranesbill with its mauve through sky blue flowers, butter-yellow Marsh Marigolds, and in late summer, the blue heads of Devil's-bit Scabious.

'We're also doing our bit to try and alleviate flooding downstream; we've blocked up a lot of old grips and planted native trees in the ghyll, fencing around the area to prevent our livestock grazing in there,' says Mr Hattan. As well as funding the tree planting, their Stewardship agreement funds grip blocking by damning up the end of each channel, paying a one-off £14.80 for each grip blocked to stop it carrying water downstream. If the upper part of a river's catchment has large numbers of these grips that are then blocked, the peak height of a flooding river carrying storm water downstream can be reduced. Blocking up a series of grips also re-wets the moorland around them, restoring the habitat to benefit its typical plants, invertebrates, and breeding birds.

So much of our upland landscape is thoroughly sheep grazed; heavy rain flows rapidly off its slopes into streams and rivers, delivering potentially massive volumes of water to lower lying land downstream in double-quick time. Millions of people in the lowlands are vulnerable to flooding, a situation that is predicted to worsen in coming years in the north and west of the country, especially as climate warming takes increasing hold. Leaving some upland areas grazing-free to develop a more rank vegetation of heather, heaths, bracken, and scrub will help prevent runoff. While these changes will benefit wildlife substantially, they will not prevent downstream flooding – the rainwater will run off eventually – but slowing its progress might well ameliorate flood peaks and reduce the damage, costs, and heartache that such flooding causes. As a societal need, the future arrangements for financial supports paid to farmers need to take account of the role modified farming practices in our uplands can play in flood alleviation. 'We wanted a traditional breed of cattle that can cope with the harsh climate and poor grazing up here and not need a lot of supplementary feed and forage as modern breeds do, yet produce milk and meat,' says Andrew Hattan. 'Northern Dairy Shorthorns were grazed here in the 1950s but went out of fashion and the breed nearly went extinct in the 1980s. We need 15 or 20 Shorthorns to milk once a day, keep them inside in winter, and make our own cheese. They produce quite a rich milk which traditionally had been used in the north to make a blue, crumbly cheese. And we're going to market it as a unique cheese that echoes these traditions and the tough upland pasture and species-rich hay meadows the Northern Dairy Shorthorn milk comes from.'

Mr Hattan is also moving away from the sheep breeds he has now – mostly Scottish Blackface and Texel crosses – and buying in the aptly named Easy Care breed. These sheep require minimal shepherding and veterinary care, shed their wool naturally in summer, don't need shearing (it isn't cost-effective here anyway), and yet offer excellent meat yields and lambing ratios. This means that Mr Hattan can spend much more of his time producing cheese and ensuring that the land is managed to the advantage of its upland wildlife.

Tara Challoner's view is that these Shorthorn cattle grazed at low intensity with sheep, together with the now more flower-rich hay meadows, the use of only organic fertilisers, and no pesticides on the farm, provide a good habitat for the typical upland wading birds that breed here. Clearing rushes and creating a patchwork of different grazing levels by rebuilding walls between pastures to control livestock numbers have also helped increase the populations of ground nesting waders. Woodcock and Common Snipe breed here, and there are good numbers of Curlew, Lapwing, Oystercatcher, and Redshank, all of which, with the exception of Oystercatchers, are declining nationally to varying extents. Golden Plover use many of the pastures for feeding (but breed on higher moorland) while Skylarks are commonplace. Black Grouse are also slowly increasing in numbers here following their recent re-introduction nearby.

UK-wide, breeding Curlew numbers declined by 16% between 1980 and 2013, though that rate of decline has slowed slightly in more recent years, their numbers falling by 3% between 2008 and 2013.[ii] The fall in breeding Common Snipe numbers has been much more dramatic: 84% between 1975 and 2013, and that decline has continued almost as steeply in the past few years. What's needed is more of Andrew Hattan's recipe on very many more upland farms across the country.

Mr Hattan adds that there's a good range of birds of prey using the farm too, a sign that the land supports a healthy population of plants, invertebrates, and up the food chain to small mammals and birds to make it worthwhile for them to hunt here. He mentions Buzzards, Sparrowhawks and Peregrines, plus Red Kites when the hay is being cut (presumably taking large ground insects), and Barn and Short-eared Owls. Funded by Countryside Stewardship (and supported by BPS subsidies) to manage this upland farm the way he does is clearly paying dividends for its wildlife. If Mr Hattan can create a decent business from producing a specialist cheese using the farm's remote location and its wildlife to promote it, Low Riggs could go from being a marginal upland farm to a viable business.

He has trenchant views on upland farming. 'Most hill farmers are old and set in their ways, not willing to innovate,' he says. 'They rely on the BPS subsidies and other payments to keep them farming the way they always have. But if the supports are reduced much more – or maybe removed altogether after Brexit – farms like ours up here aren't viable. You have to think of ways to keep farms like this going, using their location and their value for wildlife to market your locally made products, often very specialist products.'

Less haunting than ever. Eurasian Curlew (courtesy of Andreas Trepte).

129

So why do Andrew and Sally Hattan want to live in such an isolated farm along a five kilometre rough track with no neighbours in sight, a long way from any amenities, a lengthy daily trip for their children to get to school and back, and all with a high rainfall and harsh winter weather guaranteed? 'I've always wanted to farm', says Mr Hattan, 'and I spent time when I was younger on a farm in Wensleydale so I got to love the uplands. We don't feel isolated up here; we pick up other farmers' children down the Dale when we take ours to school every day; Sally works part-time as a dentist in Pateley Bridge; we have Wi-Fi and a four-wheel-drive truck so we're rarely cut off. And we love the wildlife up here,' he adds.

'Upper Nidderdale has nearly 60 farms. Most of these are in the existing agri-environment scheme, Environmental Stewardship, some at Entry Level but several at the Higher Level providing much more value for wildlife,' says Ms Challoner. 'The farmers depend on these schemes to bolster their income. The competitiveness and bureaucracy associated with the new Countryside Stewardship scheme means that it isn't a good fit for upland farms and it's a struggle to get these farmers into it. There's no guarantee they will get accepted just because they were in the previous scheme. This is already having negative consequences for habitats here; without the environmental payments farmers are forced to intensify their farming to make up the money. This puts the breeding birds of the upland grasslands and blanket bog under threat so birds including Curlew, Redshank and Lapwing will be at significant risk of further decline.'

Not far from Nidderdale, Tom Lord farms 67 hectares of upland pasture and hay meadow in the Yorkshire Dales National Park. His land is at about the same elevation as Andrew Hattan's but its thin soils on limestone drain more quickly and none of his land is waterlogged. Driving down the narrow lane to Lower Winskill Farm one mid-July day I pass off-white, lichen-encrusted limestone crags with bunches of yellow-flowering Lady's Bedstraw around them and the occasional Devil's-bit Scabious just showing its pale blue flowers.

Tom Lord has been farming Lower Winskill for 30 years and has been in some form of Stewardship scheme since the mid-1990s. He has always been keen to combine his farming with education so he's developed an interpretation centre in one of his barns and a teaching room for the many group visits he gets, most of them from schools.

'There's 14 hectares of hay meadow, 4 hectares of woodland, and the rest is pasture that I graze all year with sheep,' says Mr Lord. Ten hectares of the hay meadow, all the woodland, and 24 hectares of pasture are in Stewardship. 'Most of the lambs I sell off for fattening. In winter I borrow about 20 Highland Cattle to graze the pastures and the hay meadows and I keep the hay off a couple of fields, enough to help feed the stock through winter. We don't seem to get much snow up here in recent years but it can be pretty cold in the wind and rain in winter. None of my land has been ploughed and reseeded; I don't use any artificial fertiliser, just some manure, and no weed killers or other pesticides.'

That becomes obvious when Mr Lord takes me on a walk around the meadows and some of the pastureland close to his farm. The meadows are full of Yellow Rattle that has set seed like many of the other plants, such as the huge numbers of spring-flowering Cowslips. However, the magenta-pink heads of late summer-flowering Betony are on full show, a hay meadow plant that has declined substantially and, like many others, was once famed for its medicinal properties. A few pink Common Spotted Orchids dot the canvas too. These meadows will be cut in a week or so, the flatter ones mowed, those with a scatter of limestone outcrops have to be cut by scythe. But Tom Lord's meadows weren't always like this.

'When I came here first, all these meadows were full of pernicious thistles. I slowly restored ten hectares by cutting the thistles before they flowered and repeated it each year whenever they grew back. Gradually, that reduced their numbers. And without fertiliser, the meadow flora has returned. It's a wonderful sight to see all the colours in early summer. Some of the meadows poorer in flowers we have improved by scattering seed harvested from other hay meadows here,' says Mr Lord.

The hay meadows at Lower Winskill are recorded in late-16th century documents; they belonged to three smaller farms at that time. The farms were basically little dairy farms. Small-scale dairy farms became widespread in the Yorkshire Dales in the 16th and 17th centuries, and the milk they produced was made into cheese and butter. The development of this upland dairying economy needed ever more hay meadow to produce food for the milk cattle which were kept inside in stone built barns from early November to early May each year.

Standing amongst these meadows that Tom Lord nurtures with careful and considerate management, I find it hard to believe how rare such flower-rich upland hay meadows have become across the UK. Current estimates suggest that only 1,000 hectares of traditional upland hay meadow remain in England, and less than 100 hectares in Scotland.[iii] They are mainly in the northern Pennine valleys and parts of North Yorkshire and Cumbria. It's both the variety and abundance of grasses and flowering plants that make these meadows so valuable for wildlife, as they provide a plentiful supply of nectar for bees and habitat for a huge variety of other invertebrates.

In turn, these attract insect-eating birds like Barn Swallows and House Martins in the day, Whiskered and Noctule Bats at night. Birds including Twite, Lapwings, and Yellow Wagtails rely on these meadows for food and nesting places, and until maybe the 1950s or 1960s they were home to the now globally-threatened Corncrake. What put paid to these once-common skulking ground birds were the introduction of mechanical mowing – which cut crops earlier and faster (destroying chicks and late Corncrake nests) – and the conversion of so many hay meadows to grazed pasture, which provided no vegetation cover in which these enigmatic birds could breed.

Centuries in the making, upland hay meadows have quickly been destroyed by intensive farm management. Together with ploughing up to create fast-growing grass leys, repeated cutting for silage instead of hay has been a major cause of loss.

*Flowering gems. Flower-rich upland hay meadows at
Lower Winskill Farm, Yorkshire (courtesy of Tom Lord).*

This is along with other changes in farming practices such as increased use of herbicides and fertilisers, extensive drainage, and heavy spring grazing. A recent survey of meadows in the Yorkshire Dales National Park revealed that less than 5% could now be described as flower-rich. A similar survey in the Peak District showed that 75% of such meadows known to exist in the 1980s had either been lost or degraded by the mid-1990s.[iii]

The annual cycle of upland hay meadow management in the Dales begins in the spring, when the meadows are grazed by lambing sheep so that the first flush of grass is eaten by the ewes and lambs. This is very important as it prevents the taller grasses shading out the wildflowers that start to grow a bit later. It also encourages the growth of wild clover which naturally promotes soil fertility by fixing nitrogen in the soil. In mid-May after a few weeks of spring grazing, the ewes and lambs are taken out of the meadows which are then 'shut up' to let the vegetation grow to make into hay in July. Then they are grazed again until spring.

Lower Winskill's pastures as well as its hay meadows are covered with flowers. Lady's Bedstraw, with its frothy yellow blossom, imparts a heavy, honeyed aroma as we brush past it and stand on a small limestone outcrop to look at what's around us. Nearby are a lot of equally yellow but prostrate Common Rockrose, the sole food for the caterpillars of the very uncommon Northern Brown Argus butterfly but which is increasing at Lower Winskill; Wild Thyme flowering violet-pink on slivers of soil on the tops of rocks; a scatter of Carline Thistle, its golden-brown flowers not yet open; and many others too. Tom Lord gets paid £200 a hectare from Countryside Stewardship for the pastures; no pesticides, no fertiliser, and grazed only in winter, unlike his non-Stewardship pastures which are grazed all year (though still untreated with fertilisers and pesticides). Yet there's still a lot of maintenance on the farm; Lower Winskill's fields are small and it has over 11 kilometres of dry stone wall as a result. There's a pride in keeping them in good condition and rebuilding any sections that fall. But it all takes effort and cash.

'Three-quarters of my income comes from the BPS subsidy payment and my Stewardship scheme. It's the same for all the farms up here, even the bigger ones,' says Mr Lord. 'I wouldn't manage to survive here otherwise; the only way would be to slowly breed up a pedigree cattle herd so that I could sell them at much higher prices. I suppose I could get registered as organic because I'm virtually organic now. Or diversify even further, maybe into B&B or convert one of the outbuildings into a holiday cottage. But I'm getting a bit old to start up new things and it's the teaching I like when people and children come here to learn about farming and wildlife. I'll keep doing that.

'Nearly all the farmers up here are in Stewardship, at least on parts of their farms but no one talks about it. There's no real peer group recognition for environmental work amongst my neighbours, only about livestock. I suppose it's embarrassment; they are farmers and they seem to think receiving environmental payments isn't quite acceptable. It's a pity; it's as if they still think it's not real farming. Maybe that will change if younger people take over; we're all old, 60s plus.'

Apart from grouse shooting on large upland moorland estates in Scotland and a few in the north of England, it's grazing by sheep and beef cattle that dominates upland land use in the UK. On the higher mountain slopes, it's sheep alone. Some upland habitats require little or no grazing to retain their vegetation and characteristic wildlife; blanket bog is a classic example because most of it is so waterlogged that scrub and trees are less likely to develop. Heather moorlands are often grazed by sheep and burned in patches to eliminate old, woody heather and to regenerate new, more palatable growth. But many parts of the uplands in England and Wales – less so in Scotland – have been grazed too heavily by sheep in the past, encouraged by the system of headage-based subsidy payments available to all livestock farmers in the uplands for many years up until 2005.

Sheep grazing is especially damaging for one upland habitat of extreme wildlife and scenic significance: oak woodland. It threatens the long-term survival of the wildlife-rich remnants that are clinging on the steep slopes of many a valley in the west of Britain, in the hills of Wales and the Lake District especially. These so-called 'oak hangers' are amazing places; their often weather-beaten trees – mainly oak but with some ash, rowan, hazel, birch, and holly – can be laden with a plethora of lichens and mosses adorning their bark, benefitting from a high rainfall and clean air. They are the breeding grounds of Buzzards, of chessboard-coloured Pied

Flycatchers and resplendent Redstarts that fly here each year from central Africa, as well as tiny Wood Warblers with their trilling songs. Polecats roam these woods at night, and maybe there is even an occasional Pine Marten, while Red Foxes and badgers are plentiful.

Hill farmers value these places as shelter for sheep in harsh winter weather, although ironically their frequent grazing means the eventual end for an oak hanger. Seedling trees trying to grow – the new trees needed to replace the denizens that ultimately die – are chewed off, never making it even as far as saplings. So farmers need to be encouraged to fence off much of their hillside wooded land to keep sheep out, at least on a rotational basis, thereby allowing new trees to get well above grazing height before sheep are let in again. It's a measure included in all of the UK's agri-environment schemes that must continue and needs to be part of many more farmers' agreements.

MUCH LESS HAUNTING

The haunting sound of the Curlew's display call ('Cur-lee') echoing across the moors in a swirl of mist epitomises the wildness of much of our uplands and is unmistakable. Heard from February through to July on its breeding grounds – mainly moors, boggy heaths, and upland pasture – in winter, the birds move to the coast, especially our large estuaries, where British birds are joined by large numbers seeking shelter from the harsh winters of northern Europe.

But the Curlew's haunting calls are not as common a sound as they used to be. Curlews have declined mostly in the west of the UK, in Scotland, Wales and in South West England; even more so in Northern Ireland and the Republic of Ireland. The highest breeding numbers are in northern England, especially in the Pennines, eastern Scotland, and the Northern Isles, hence their frequent calls echoing across the lonely moors around Low Riggs farm. Britain holds one-third of their world breeding population but they declined by 19% between 1980 and 2014.[ii]

Destruction of upland wet grasslands and moors by planting them up with conifers, draining parts of them, and even ploughing up small areas; conversion of rough grassland to grass leys or to grow arable crops; increased predation by foxes and crows; and more intensive upland farming practices including keeping higher numbers of livestock have all contributed to this emblematic bird's decline.

Too-frequent burning of vegetation is another problem for much hill land; a combination of heavy sheep grazing and excessive vegetation burning has put paid to large areas of heather-dominated upland and replaced it with much more plant and animal poor coarse grassland. Where once Ring Ouzels (mountain blackbirds), Golden Plovers, and Merlins might have nested, there are now huge numbers of Meadow Pipits and little else. Plant richness has declined; so, too, has the variety of invertebrates and small mammals. In large tracts of Central Wales, coarse grasses now dominate vast upland areas that support a meagre assemblage of invertebrates and little variety of upland birds. Today, sheep numbers in the uplands have fallen, although much of our mountain and upland is still subject to too much grazing and too much burning.

Incomes on the UK's upland sheep farms are consistently much lower than on lowland farms; the lifestyle is very much harder and more prone to major weather-related emergencies; there are fewer opportunities to diversify the business; and the infrastructure of roads, electricity, broadband access, and other needs are often restricted. However, upland livestock farming doesn't exist in isolation; it is inextricably intertwined with lowland farming. For instance, most lambs born on hill farms are kept for several weeks then sold off to lowland farms for fattening on better grazing before again being sold for slaughter. It isn't possible to raise and fatten lambs ready for eating on most hill land.

John Lloyd, who keeps a pedigree flock of 250 North Country Cheviots on about 49 hectares of upland summer grazing near Llandovery in Central Wales on his 142 hectare farm, is convinced that most hill farmers – unless they have other significant sources of income – would struggle to survive without their BPS subsidies.

'Sheep numbers generally are falling in the hills since the headage payments were done away with years ago,' says Mr Lloyd. 'Less available labour leads to gathering in difficult terrain becoming unsustainable; if one grazier gives up on one side of a mountain area, shepherding on the rest of the common becomes more difficult as sheep wander and there are fewer shepherds these days to share the work. Some will probably give up and sell out to incomers who won't bother to graze much of the hill land, only the pastures near to the farm. Others will sell to forest interests, particularly now it has become easier to get the relevant permissions to plant areas of open hill with conifers. Or they'll let out the land to sheep ranchers who have no interest in looking after the land but tow mobile sheep handling units from place to place; they lamb outdoors and sell them off for fattening. They don't usually overgraze the land overall but it can be heavily grazed at some times of the year. A few farmers around here already do that,' he adds. All in all it's a haphazard approach to changes in land management with unknown consequences for wildlife.

He agrees that some parts of our uplands and mountains that are now hardly grazed at all – the steeper slopes and areas that are less accessible – are reverting to a more wildlife-rich set of habitats in which hawthorns and rowans (and on occasion unwanted Sitka Spruces from nearby plantations) are slowly rising through increasingly dense patches of bracken and heath. These habitats are replacing heavily-grazed grasslands, thereby improving the habitat for wildlife. It's a significant gain for a wide range of plants and animals, including breeding

Cuckoos, Whinchats, Tree Pipits, Linnets, and many more. However, their falling out of farming use does mean that particular species such as Wheatears and inland-breeding Chough are already declining because they inhabit more closely grazed upland grasslands. Gathering sheep on rocky, steep slopes like many upland valley sides to get the sheep down to lower ground in autumn, or for other needs such as dipping against sheep parasites, is hazardous, disproportionately time-consuming, and often not worth the effort. In years past, hill farmers would walk or go on horseback with their dogs to gather sheep, covering every bit of land; nowadays, they go where their quad bikes will go and that doesn't include steep, rock-strewn slopes!

Other upland areas where sheep numbers have fallen significantly in recent years have been subject to heavy sheep grazing and vegetation burning over many years (sometimes centuries). These areas have now developed dense, tussocky, Purple Moor Grass-dominated vegetation that's poor for wildlife and virtually impossible to walk through. Very few other plants can compete with this grass, one of the toughies of the plant world. Its invertebrate populations are limited, and Meadow Pipits and a few Skylarks are about the only birds that breed in it. Diversifying such dense sweeps of this tough grass is almost impossible.

John Lloyd suggests that heavy grazing by beef cattle would help break up its dominance as it does on similar ground in Dartmoor but agrees that keeping cattle in open areas when there is no recent history of such management is fraught with difficulties. Regular burning of the Purple Moor Grass would actually increase its density because it encourages a plethora of young shoots to regenerate each spring. Other possible alternatives, such as dropping bales of heath species from helicopters to try to get their seed to germinate, or using mechanical equipment to increase the areas of bogs to inundate it with water and deter its growth, would cost money that doesn't seem to be available. That leaves only one other option: planting a scatter of broadleaved trees such as willows and others that might cope with almost year-round waterlogged conditions and slowly shade it out. It's an option only likely to be taken up where such land is owned by conservation organisations.

But in spite of being modified by heavy grazing, vegetation burning, and more piecemeal bits of drainage and habitat clearance, the majority of the farmed uplands retain considerable areas of habitat, unlike most of the farmed British lowlands where much of the wildlife habitat has been obliterated or disassembled into tiny, often scattered remnants. As Dr Lois Mansfield, an upland farming expert at the University of Cumbria argues:

> It becomes apparent that we need a delicate symbiosis between farming and ecological systems if both are to be sustained in the uplands. For this to occur, upland farm businesses need to be economically viable without over-exploiting the environment, something which has proved difficult to achieve.[iv]

We need to retain extensive areas of heather moorland and blanket bog as open, largely treeless habitats because they have developed their rich fauna and flora over very many centuries. Light sheep grazing will do that. But we also need to reduce grazing levels on many upland slopes so that they slowly transform themselves from a billiard table-like turf of very little value to wildlife back to a better-developed mosaic of vegetation, including some heath, bracken, and areas of scrubby broadleaved woodland. That will be a substantial boost to wildlife and will also help slow the down flow of rainwater and ameliorate the peaks of river-flooding downstream that might become more commonplace as a result of climate warming. It can be achieved only by drastically reducing grazing levels or by curtailing all livestock grazing in some places.

The Pontbren Project, named after the upper reaches of a tributary of the River Severn in Mid-Wales, is an innovative approach to using woodland management, tree planting, and no grazing areas to improve the efficiency of upland livestock farming led by a group of neighbouring farmers.[v] They developed new on-farm uses for woodland products, and when it became clear that tree planting had not just improved farm businesses and wildlife habitats, but had also reduced water run-off during heavy rain, they invited scientists to investigate. Research revealed that strategically-located belts of trees and areas of vegetation left ungrazed are effective at reducing the rate at which water runs off hill slopes. Wildlife has benefitted too, as a variety of plants and shrubs such as heather and bilberry plus a few trees associated with our uplands – birch, rowan, and hawthorn especially – gained a foothold. Unusually, the Pontbren farmers decided not to make use of any existing agri-environment scheme and its payments; they decided that the scheme options available were too rigid, not containing enough flexibility for what they wanted to achieve.

A brilliant example of the impact of removing sheep grazing is visible at the Cwm Idwal National Nature Reserve in Snowdonia. Sheep had grazed this cwm (or corrie), famous as being the location where Charles Darwin worked out glacial theory, for centuries. Apart from rock and scree, the whole of it was vegetated with closely-grazed grassland. With the agreement of the tenant farmer, sheep have been excluded since 1998. Even though stray sheep entering the Reserve from adjacent mountain land (it is mostly not fenced nor actively shepherded) are a continuing problem, visually the land has changed dramatically. Much of it is now dominated by rising amounts of Common Heather rather than grassland, with increases in the frequency of Bell Heather on better drained ground and Cross-leaved Heath and Crowberry in more poorly drained and waterlogged areas. Several other flowering plants have also increased in abundance and it's very likely that invertebrate populations have increased, though there is no monitoring yet to prove this.

Little wonder then that the Green Alliance – a charity and independent think tank focused on ambitious leadership for the environment – is calling for a post-Brexit policy that rewards farmers for adopting natural flood management such

The difference a fence makes: heavy upland sheep grazing, left; ungrazed right.

as tree planting, reinstating water meadows (that hold flood water), blocking moorland drains, building timber blockages in rivers, and recreating river meanders to slow flows.[vi] It calls for investment to be taken away from subsidies and reinvested in flood-prevention land management. Elements of agri-environment schemes that focus on reducing livestock grazing levels in the uplands, fencing off steep slopes to encourage scrub and rank vegetation to develop, and generally decreasing the vast areas of billiard table grassy slopes all help slow peak flood flows and need much more cash and effort devoted to them.

No simple adjustment in the equivalent of any BPS-like payment given as a subsidy to farmers post Brexit is going to achieve a sufficient balance between retaining hill farming and improving the upland environment for wildlife. Removing the subsidy payment completely would put many upland farms, especially the most remote and the smallest, out of business, and risks ushering in large-scale sheep ranching with little or no care for any environmental issues, wildlife included, or extensive conifer plantations of very limited wildlife value. If sheep numbers in the uplands fall further – combined with more farmers deciding that it's not worth the effort to gather in stock on some of the steepest and more inaccessible terrain – further small areas of upland, steep valley sides in particular, are likely to re-develop the lush vegetation that once carpeted them.

Retaining subsidies for hill farms but with a realistic cap on the maximum paid to any one holding, together with making agri-environment agreements compulsory on all farms, could offer a sensible way ahead retaining the majority of upland in agricultural use while boosting its wildlife value. Future agri-environment schemes need to focus largely on further reducing grazing levels overall and on removing some tracts of land entirely from grazing by fencing them out. These measures should be seen as beneficial for wildlife and of societal importance because they contribute to the decline of downstream flooding.

Endnotes

i   Farming in the Uplands. Written evidence submitted by Defra (Uplands 18) to Parliament, Session 2010–11.

ii 'Wild Bird Populations in the UK, 1970 to 2014,' Annual statistical release. Defra, 2015.

iii 'Upland hay meadows,' The Wildlife Trusts, 2015.

iv *Upland Agriculture and the Environment,* by Lois Mansfield. Badger Press, 2011.

v   'The Pontbren Project: The Farmers' Experiences & Lessons Learnt,' Wales Rural Observatory, 2013.

vi 'Smarter flood risk management in England: Investing in resilient catchments,' Nicola Wheeler et al. The Green Alliance, 2016.

CHAPTER 10

BACK TO ORGANIC

We only invented the word organic because we made things inorganic.

KHANG KIJARRO NGUYEN, ARTIST, PHOTOGRAPHER, AND PERFORMER

High above the village of Malham in the Yorkshire Dales National Park, along several kilometres of stony single-track road on land owned by the National Trust, is the somewhat-euphemistically named and very isolated New House Farm. In winter it's often a harsh, windswept, and very wet spot. But visit on a bright summer's day and you will be astonished by the assemblage of tiny specks of seemingly every colour and shade imaginable in the surrounding hay meadows. New House Farm, tenanted by Roy and Irene Newhouse (their farm-matching surname is mere serendipity) has arguably the richest wildflower meadows in the Yorkshire Dales.

Their farm has been registered as organic since 2006; its 28 hectares support 14 beef cattle (Blue Greys) and 50 Lleyn pedigree sheep. 'We've got a mixture of pasture we graze all year and about 11 hectares of hay meadows, all of it in the Stewardship scheme, so we don't cut the meadows until after 15 July any year. We spread a bit of cattle manure on the meadows but very little on the pasture and we're self sufficient in hay for feeding the stock in winter. There's 35 different flowers and grasses per square metre in the meadows,' says Roy Newhouse.

They're a mix of Sweet Vernal-grass (it has a sweet vanilla taste if chewed); mauve-blue Wood Cranesbill; foamy Ladies Mantle; Pignut, with its umbels of tiny white flowers; masses of yellow-flowering Hay Rattle; and stately Melancholy Thistle, purple flowering but spineless, and once used to cure melancholia as its name might suggest. The limestone outcrops in some of the pastures here are home to Alpine Cinquefoil with its dazzling yellow flowers, and Orpine, a purple flowering, long-stemmed sedum. Not surprisingly, this cornucopia of flowering plants is home to a huge range of birds and insects. 'We've got Common Snipe breeding up here and there's some Curlew, Lapwing, and Oystercatcher. We usually hear Redshank but we haven't this year. There were a few Oystercatchers perched on a wall just over here a bit ago before you came. They're always calling loudly,' says Mr Newhouse.

As at New House Farm, organic farming shuns the use of artificial pesticides, fertilisers, and other agro-chemicals; it doesn't use GM crops or other advances in farming biotechnology; it rejects the routine use of pharmaceuticals on livestock; it promotes high standards of animal welfare; and it emphasises crop rotation, soil quality, and maintaining biological diversity as alternatives to chemicals. In this way, organic farming is, in many respects, a return to 'traditional' farming methods.

Organic Pros and Cons

Positive aspects of organic farming:

- The environment benefits because natural habitats are better conserved
- The soil can be in better condition because of the manure used
- It can provide healthier food for people
- Fewer chemicals that harm bees and other insects are used
- The industry is worth over £1 billion a year in the UK

Negative aspects of organic farming:

- More produce is damaged by pests
- Weed control is time consuming as weeds are often removed mechanically
- Some organic pesticides, such as copper, can remain in the soil and be harmful
- Organic dairy farms produce more methane per animal than non-organic farms
- Some organic farming methods use more water than non-organic methods
- The crop yield is lower on organic farms
- Most of the organic food bought in the UK is imported

(Source: www.bbc.co.uk/.../rural_environments/farming_rural_areas_rev4.shtml.)

In the UK, products that are produced in line with organic standards have to be labelled as such. There are nine different organic certification bodies in the UK, including those run by the largest, the Soil Association (SA) and Organic Farmers and Growers (OF&G). Any organic farming scheme is required to comply with standards set by the Advisory Committee on Organic Standards which in turn ensures compliance with European and international requirements. By law, products labelled as organic must display a certification number or symbol. Post Brexit, compliance with a European standard will presumably not be mandatory, though it will doubtless still be essential for any organic product being sold in EU markets.

According to principles laid out by the Organic Research Centre (ORC) established in 1980 and overseen by the Progressive Farming Trust Ltd., the Centre's parent educational charity, organic farmers aim as far as possible to have few if any external inputs on their farm and keep waste to a minimum.[i] Consequently, recycling is important and manure from livestock is treated as a valuable resource. Soil is considered to be of vital significance and maintaining its long-term fertility is a priority. Biologically-active soil will decompose organic matter faster, encouraging bacteria, fungi, and earthworms; when soil organisms decompose organic matter, nutrients essential for plant growth are recycled back into the soil for the next crop. It's important, too, for the soil not to be compacted by livestock in wet winter weather or by heavy farm machinery.

The ORC says that avoiding pollution is also a main concern, meaning good waste management and farming practices are essential, as is keeping the use of fossil fuel in food production to an absolute minimum. They cite that between four and six tonnes of crude oil is needed to make just one tonne of synthetic fertiliser. Manure produced by farm livestock can be very polluting, so organic farmers use it as a natural fertiliser on pasture but only in spring or summer because it will decompose quickly then and help plants grow. It is never spread in winter when the soil organisms are too cold to decompose it and the crops are too cold to grow; it could be washed off by heavy rain and seriously pollute streams and rivers, depleting their wildlife.

Organic standards for animal welfare are very strict. Livestock must have access to food and water, with outdoor grazing wherever possible. They must have plenty of space in fields and indoors in winter. Organic farmers are prohibited from routine use of veterinary medicines such as antibiotics, although if an animal becomes sick they can be used after seeking permission.

Supporters of organic farming acknowledge that its production methods are inherently more costly yet organic produce commands higher prices. Critics claim that organic food is an exclusive luxury, insofar as it is more expensive to buy than other food and is frequently less readily available in retail outlets in more deprived urban areas. However, according to Ray Keatinge, Head of Animal Science at the Agriculture and Horticulture Development Board, the average yield of milk per cow on an organic farm is not much less than that on a non-organic farm. Plus, organically-produced milk has a higher solids content that gives it a richer taste. It's only the stocking rate, around 70% to 80% lower on an organic farm, which makes the overall productivity lower, although not the profit margins due to the premium price organic milk attracts.

Laurence and Tom Harris, the father and son team who run Daioni (it means 'goodness' in Welsh), the organic milk company based at Ffosyficer farm near Boncath in Pembrokeshire, have been registered organic since 2002. The farm started in the 1970s with 69 hectares of mostly grazing pasture. Today, they are certified by the Soil Association and have a dairy unit of 648 hectares, two-thirds of it rented and one-third owned.

'Our decision to go organic was 75% economic: a better price for our milk; a better market for a range of organic milk products; and better for the long term sustainability of the farm,' says Tom Harris. 'The other 25% of our decision was that we didn't want to go intensive. We stock at the rate of around two cows per hectare on average and produce about two million litres of milk annually from a herd of 350 cows. We produce the Daioni product range of organic fresh milk, organic UHT/long-life plain milk, long-life flavoured organic milk, organic cream, and we recently launched "Daionic", a high-protein sports-focused organic milk drink.'

After a long chat in the farm kitchen, we walk from the Ffosyficer farmhouse to a large field holding maybe 50 cows; the lush vegetation is a mix of ryegrass, clover, Timothy grass, and Cocksfoot grass planted as a ley. Most of the fields here are on a six year rotation in which they are put over from pasture to a root crop such as turnips or forage rape followed by cereals (barley or oats) and back to grass. Some other fields on the farm are permanent pastures and these are never ploughed. The dairy cattle – a mix of breeds including Holsteins – are kept indoors in winter but out on the fields nearer the farm during the rest of the year so that they can be easily gathered twice a day for milking. More distant fields on the farm are sheep grazed.

The whole of Ffosyficer is in a Glastir agri-environment scheme which includes a variety of wildlife-friendly provisions such as fencing out livestock from farm woodland, managing streamsides to reduce livestock damage, establishing new hedges, and trimming existing hedges on a three year cycle. This is, of course, in addition to the general wildlife gains of not having any pesticide spraying or synthetic fertiliser runoff into streams. Tom Harris points out that all manure slurry on the farm is injected into the top layers of the soil on Ffosyficer's fields rather than sprayed over the surface because spraying releases some of the nitrogen as ammonia into the atmosphere (thereby losing it as valuable fertiliser and adding to air pollution) and runs more risk of runoff into streams.

The Harris's have clearly been entrepreneurial in identifying and exploiting a niche market for organic products such as long-life chocolate, banana, and strawberry flavoured milk in a variety of pack sizes. It markets them in the UK and in China, Hong Kong, and the United Arab Emirates, and has won several awards over the years. A detailed financial report on organic farming in England and

*No synthetic fertilisers, no pesticides. An upland hay meadow farmed organically
in the Yorkshire Dales brimming with late summer flowering Lady's Bedstraw.*

Organic milk. Tom Harris with some of his organic dairy cattle, Ffosyficer farm, South West Wales.

Wales for 2014/15, undertaken by the ORC for the Welsh Government, showed that the organic dairy industry is now generating higher profits than non-organic farms despite producing lower milk yields.[ii] That's mostly because its fertiliser costs are much less and it attracts a premium product price.

With organically-farmed land representing just 3% of the UK's total farmed area, the UK Government has come under criticism for allegedly taking a half-hearted approach to promoting organic farming, while many other European governments have acted more confidently.[iii] Germany, for instance, has over 23,000 organic producers farming well over a million hectares; France has much the same; Italy has 44,000 producers farming 1.2 million hectares while the UK has around 520,000 hectares (58% of it in England) and around 3,500 producers, down from over 5,000 in 2008.[iv,v] Denmark is the world leader with 7% of its farmed land registered as organic, and it's still increasing.

In every European country except the UK and Norway, the land area farmed organically increased between 2010 and 2015. In the UK it has been declining each year from a peak in 2010 of 700,000 hectares to less than 500,000 hectares in 2015. There might be a tiny glimmer of more positive news; the area of farmland in the UK being converted to organic (the process usually takes two years) has been falling since 2007 but showed a slight up-swing in 2015. Between 80% and 90% of organic farmland is pasture for grazing livestock followed by cereals and vegetable growing as the next most important organic products.[iii] The numbers of livestock farmed organically are all in decline, though organically-raised poultry with two and a half million birds (out of a UK total of nearly one billion birds!) are just holding their own.

Ironically, though, sales of organic produce are rising. In 2015 they went up by 4.9%, outperforming the wider food and drinks market where sales dropped by 0.9%.[vi] The figures suggest the organic market is continuing to recover after a short-lived decline between 2008 and 2011. But paradoxically, much of the organic produce on supermarket and grocers' shelves is imported! When the financial crisis took hold in 2008, organic foods were among the first to take the hit as consumers cut back on their weekly shopping bills. At the time, almost half of organic shoppers said they would reduce or even give up buying organic food in the following year.

'Some of the fall in the number of organic farmers and the land area farmed organically is due to the financial downturn we've gone through since 2008 and we've still not recovered,' argues Steven Jacobs, OF&G Business Development Manager. 'Supermarkets removed lots of organic produce from their shelves because customers often didn't want to pay premium prices. But smaller retailers are still showing healthy growth in organic products and two of the big discounters, Aldi and Lidl, are stocking more. I think it will pick up but you can't just turn it on suddenly because the whole supply chain needs to be in place, from the farmer through processors to wholesalers and retailers. It needs support from field to fork. I don't believe there's enough support for organic farming in the UK; Defra, for instance, could be more positive. I think it needs more investment, helping organic food processors for example and not just organic farmers.'

The Soil Association claims that the increasing consumer demand for organic products has still not hugely impacted organic farming in the UK, although Defra's report for the last year does show a slight increase for land in conversion to organic in the UK, the first increase since 2007.[iii] 'It's surprising,' says Peter Melchett, the Association's Policy Director, 'that more UK farmers are not responding to demand more quickly. We still have a long way to go to reach the potential we see being realised in other countries such as Denmark where there is a 30% market penetration for organic milk. There are good markets for organic cereals in the UK, especially for bread-making wheat, but not enough organic wheat is grown here in the UK.'

Some organic farmers who have gone back to non-organic production mention issues such as the bureaucracy involved and the rigid standards that must be adhered to in order to convert and to retain organic status year on year as well as problems overcoming persistent weeds because no herbicides can be used. The Soil Association doesn't accept that these are significant issues and Tom Harris comments that he has found no such problem in his several years of being registered organic.

'We are always working to make life easier for our licensees,' says Mr Melchett. 'Whether that means bringing in new online systems for monitoring or adopting best practice in terms of risk based inspection which does reduce the bureaucracy and time involved. But we must assure consumers and producers that organic food meets the legislative requirements so we do need to check carefully. The legislative requirements behind organic standards mean that you can trust in organic food. Regarding weeds, the opposite is actually true! Some farmers are choosing to go organic due to herbicide resistance on some non-organic farms, especially to overcome pernicious weeds such as Blackgrass. Research is taking place to look at new ways to control them such as electric weeders and use of deep surface cultivations for creeping thistles.'

'We've learnt from other European countries that what's needed is much more government encouragement for organic farming. We set out this case in a report we wrote some years ago but the arguments and case studies still hold true. We said then: "We have found that most European countries have acted confidently to normalise and champion organic food and farming as a pioneering, sustainable and environmentally friendly way to produce food. In contrast, UK governments have been diffident, if not lazy on the subject."[i] The high proportion of organic food required in publicly-funded cafes and canteens, for example in schools and hospitals, in countries like Sweden and Denmark does not generally involve much or any extra cost (menus can be adjusted to keep prices steady), and do make eating organic food normal, something that everyone does,' he adds.

Controversy rages about whether organically-produced food is healthier to eat than produce from non-organic farms (called 'conventional' by farmers even though it's only been 'the convention' for less than a century!), a debate that focuses on pesticide residues, GM products, animal welfare standards, food nutritional values, and, to a lesser extent, food contamination with pathogens. But it's a debate that is not relevant to the subject of this book.

AN ORGANIC HISTORY

The term 'organic farming' was coined by Lord Northbourne (1896–1982), in his book, *Look to the Land* (J. M. Dent, 1940). The movement has its intellectual roots in European theories of 'biodynamic agriculture' developed by Rudolph Steiner; the work of Sir Albert Howard in the 1920s; and examples from its practice in 1930s Switzerland. Lady Eve Balfour and others who were similarly minded founded the Soil Association in 1946, postulating that there were direct connections between farming and plant, animal, human, and environmental health. For its first 30 years, the SA was based on a farm in Suffolk where research helped to shape the SA's first organic standards in 1967.

Through the 1950s the adverse effects of 'modern' farming continued to kindle a small but growing organic movement. Other events boosted the movement, including the publication of Rachel Carson's influential *Silent Spring* (Houghton Mifflin, 1962) which detailed the damaging impact of DDT and other chemicals on the environment. The social unrest of the late 1960s, leading up to the oil crisis of 1973, stimulated the 'back to nature' philosophy of the modern environmental movement. Organic farming in particular condemned the environmental degradation inflicted by the Common Agricultural Policy (CAP).

In the 1980s, farming and consumer groups around the world began seriously pressuring for government regulation of organic production. In 1984, Oregon Tilth established an early organic certification service in the US. In 1991, the EU passed Council Regulation EEC 2092/91, which set the standards that all EU organic producers are required to meet.

In 2002, the UK Government produced an Organic Action Plan as part of its Strategy for Sustainable Farming and Food, drawn up in response to the 2001 foot-and-mouth disease epidemic. Environmental concerns and a series of food safety scares such as BSE in cattle have stimulated demand for organic produce in recent years.

So how does a farmer go about getting his farm registered as organic? The first step is to apply to register with one of the nine UK organic certification bodies and to establish a conversion plan showing how he will change his farming systems to become organic from however he farms at present. It has to include details about the cropping history of each field and a livestock plan. An inspector appointed by the certification body will then produce a report based on a farm visit; this will set out conditions that have to be met to bring the farm in line with organic standards before it can be registered. Once that's done, a certificate proving that the farm is organic or in the process of conversion is issued. It usually takes two years; after that, all organic farms are inspected annually to ensure that the standards are adhered to.

The financial risk of converting to organic production is substantially offset for a farmer by generous per-hectare payments from the UK's governments for land he converts from non-organic management. In Wales, Glastir Organic (part of its agri-environment climate scheme) supports organic conversion (up to a maximum of 400 hectares of land on any one farm) but doesn't guarantee that all applicants will be supported; that depends on how much positive environmental land management is included in the farmer's proposals.[vii] A successful entrant into Glastir Organic receives £130 per hectare for all pasture or arable land under conversion for the first two years, reducing to £65 per hectare for the next three years. In theory, after the first five-year agreement ends, an organic farmer can apply again to Glastir Organic and – if accepted into the scheme – continue to receive £65 per hectare for another five years. It's a generous package.

'We designed Glastir Organic to try and stimulate more uptake from the arable, horticulture, and dairy farming industries as opposed to the meat sector where there has been a surfeit of supply and a negligible price premium. Currently, we have 68,439 hectares of land in Glastir Organic in Wales,' comments Dr Kevin Austin, Deputy Head of the Agriculture, Sustainability and Development Division in the Welsh Government, speaking in 2016.

For most unenclosed land including hill and mountain land, heathland, or saltmarsh, the payment is much lower at £15 per hectare. This reflects the fact that almost no alteration in farm management is required because these are areas of open livestock grazing where most materials banned on organic farms – synthetic fertilisers and pesticides for example – are not used anyway. Furthermore, an organic farmer (and one converting to organic) can claim up to £500 annually to offset the cost of certification charged to the farmer by the certification body. These payments are in addition to the Basic Payment Scheme (BPS) and greening payments a farmer receives. Land in organic conversion or certified as farmed organically is exempt from meeting the greening requirements, although it still receives the greening payment.

Once in a certified organic scheme, any farmer can enter other components of an agri-environment scheme to enhance the wildlife value of a now organically-managed piece of land. 'Glastir Organic has actually been costed in such a way as to enable participation with other area-based schemes,' says Mr Austin. 'We've incorporated deductions for potential double funding so that farmers can participate in multiple schemes at any time and not just after their organic supports end.'

England's Countryside Stewardship scheme (CS), for instance, includes Mid Tier options such as 'OT2: Organic land management: unimproved permanent grassland', which pays £20 per hectare per year for pasture or rough grazing that is relatively rich in flowering plants to be maintained as such. An organic farmer is precluded from using synthetic fertilisers and pesticides in any case. Option 'OT2: Organic land management: improved permanent grassland' pays £40 per hectare on existing 'improved' grassland (poor in plant species and usually dominated by ryegrass and clover) in the hope that its plant diversity will steadily increase because no fertiliser and pesticides are being used. It could be argued that these options – and there are several more – are using taxpayers' money to pay extra for something that should occur naturally under an organic farming regime, the start-up costs of which have already been paid for with public money.

Both the Scotland and Northern Ireland schemes have similar arrangements; farmers converting to organic are eligible to apply for many, though not necessarily all, measures included in the relevant agri-environment climate scheme.[viii,ix]

In spite of these seemingly generous incentives, the SA has argued that the UK's organic farmers have been put at a serious disadvantage compared with their European counterparts because they receive the lowest organic support payments in the EU.[x] They claim that the now-growing UK consumer organic market means that more organic produce is likely to have to be imported in the near future. They are pressing the UK Government to ensure that British farmers receive payments for the multiple environmental and social benefits delivered by organic farming to at least match the average of those paid to organic farmers in all other member states. Post Brexit of course, the governments in the UK can fix whatever supports they deem necessary. Yet there is a strong case, if not on human health and nutrition grounds then most certainly in terms of wildlife gains, for our governments to promote organic farming rather more vigorously than they do at present.

It isn't essential to convert to organic farming in order to bring about considerable gains for farm wildlife. LEAF (Linking Environment and Farming) is the leading organisation promoting sustainable agriculture, food, and farming worldwide. It operates internationally and with around 500 registered farmers in Britain who have attained the necessary standard (independently verified) to display the LEAF marque on their produce. LEAF farmers produce good food with care and to high environmental requirements.

'A third of all fruit and veg sold in the UK is now LEAF certified,' Caroline Drummond, LEAF's CEO tells me. 'Some of our registered farmers have part of their farm certified as organic but most don't. LEAF farmers use pesticides and fertilisers but, and it's a big but, in the right place, in the right quantities and at the right time so we never shun modern technology. But our farmers go to significant lengths to encourage wildlife too.'

On its website, LEAF 'promotes an integrated, whole farm approach that combines the best of traditional methods with beneficial modern technologies to achieve high productivity with a low environmental impact. A recent survey carried out among LEAF members demonstrated significant cost savings because of better soil management, the use of minimum tillage (sowing crops without ploughing the soil deeply) and reduced pesticide use alongside improved wildlife numbers and reduced carbon dioxide emissions.'[xi]

This is precisely the approach being taken by Robert Kynaston at Great Wollaston Farm near Shrewsbury. His is a LEAF demonstration farm, encouraging interested groups of people to visit and learn about more sustainable approaches to farm management. Mr Kynaston farms 97 hectares including 40 hectares of arable crops and 45 hectares of pasture; the rest is woodland.

'I've been in LEAF since 1998. I was already interested in farming more sustainably and with care for the environment and what I read about LEAF seemed to offer what I wanted. It sort of fitted the bill,' he says. 'LEAF makes me think more thoroughly because they suggest a range of options that I can choose and they encourage you to go about things more sustainably and in a more environmentally sensitive way. They've made me look at areas of farming where I was weak as a farmer. That's good; it's a challenge. I really didn't like the idea of converting to organic because it's too rule based and inflexible. As a LEAF farm I encourage visits; maybe around 15 a year and I also take students from Harper Adams University (which specialises in teaching wide aspects of the rural economy) and school visits.'

Mr Kynaston milks 80 dairy cows, Holstein and Friesian crosses, selling the milk to Arla locally (a farmer-owned international milk company), with some of it going for cheese making. Almost all the cattle feed for the cows, which are kept indoors in winter, is grown on the farm, reducing costs and transport substantially. When I visited him to talk about his farming system he was loading harvested wheat into a silo but had not yet cut fields of peas and barley sown amongst a ryegrass and clover ley which he uses for silage. Although he hasn't tried herbal leys he has grown chicory with grass and clover, harvesting it to make good silage. Mr Kynaston doesn't do any deep ploughing; he's much keener on minimum tillage, although he does shallow plough in spring.

He uses pesticides on his crops and some fungicides, but he's concerned about the impact of fungicides on naturally-occurring fungi in the soil and tries to limit their use. He has created untilled, fertiliser and pesticide-free, six-metre-wide strips around his arable crops for the flowering plants, insects, and small mammals these support. They're full of flowers like Common Knapweed, Oxeye Daisy, and Yellow Meadow Vetchling, 'nothing unusual' as Mr Kynaston claims yet valuable for plenty of invertebrates nonetheless.

'I don't cut these margins,' he says. 'I just let them die back naturally in autumn. I sometimes have to spot treat any pernicious weeds such as docks and thistles with a herbicide to keep those under control. When an arable field goes into temporary pasture on my crop rotation, the cattle graze the margins down but they're full of flowers again when the field goes back to arable. All my fields except a few permanent pastures are on rotation so there are flower margins at different stages all round the farm at any one time.'

Mr Kynaston trims his hedges once every three years and he leaves wheat and barley stubbles to overwinter in some fields each year. 'We've got Brown Hares on the farm and several pairs of Yellowhammers breeding, along with Tree Sparrows and Skylarks. Lapwings nest in some fields of spring crops and then move their chicks into the cattle-grazed pastures to feed on insects. We've plenty of Buzzards around and we see a Red Kite occasionally too.'

He has been in some form of Stewardship scheme for 20 years but his latest agreement ended in January 2017. He isn't guaranteed to get into the replacement CS scheme, potentially jeopardising what he has achieved through participating in Stewardship for so long. Uncertainty about future

A new leaf? Robert Kynaston at Great Wollaston Farm near Shrewsbury is a LEAF (Linking Environment and Farming) farmer.

farm funding after Brexit, compounded by changes to the agri-environment schemes being administered in each UK country, puts at risk much good work already done on farms around the UK. It could also waste a great deal of taxpayers' money if farmers decide that without the money paid to them, they will not voluntarily retain habitats they have created or long managed under an existing scheme. Smaller farmers might not be able to afford to do so; larger lowland farmers in receipt of substantial BPS payments certainly should be able to.

Changing arrangements for agri-environment schemes is problematic; what's needed after Brexit is scheme stability in order to make sure that wildlife benefits long term. That doesn't mean that components of a scheme won't change – there needs to be flexibility – but the overall structure should remain more or less constant.

'I'm hoping to get a Stewardship agreement in Natural England's Mid Tier scheme,' Mr Kynaston says. 'If not, I'll keep some of the better margins, those next to ditches and woodland probably because they're the best for wildlife and I need to leave some margin there for farming purposes anyway. And I'll keep cutting the hedges every three years. But maybe half of the margins will have to go.'

A somewhat similar scheme to LEAF is 'Fair to Nature: Conservation Grade', which has the rather grandiose aim of 'rebuilding the world's wildlife through nature friendly production methods'.[xii] Accredited brands using predominantly 'Fair to Nature' ingredients carry a distinctive yellow 'Fair to Nature' bee logo on their packaging, leaflets, and websites. Conservation Grade is an independent accreditation organisation that facilitates commercial relationships between farmers and consumers in the UK through a system of agriculture that optimises both crop yields and wildlife conservation. They currently have just 64 farmer members (all lowland farmers and all in England) farming about 25,000 hectares and a waiting list of farmers wishing to join once they have more contracts available for the products they produce. At present Conservation Grade is very much orientated towards arable farmers although they are developing protocols for livestock farms too.

Conservation Grade farms have to create specific wildlife habitats on at least 10% of their land. These habitats include: pollen and nectar habitats created from wildflower seed mixes; wild bird food crops; tussocky and/or fine grass mixtures that are very important for small mammals, birds, and overwintering invertebrates; plus others such as hedges, ponds, and woodland. Sometimes a whole field is devoted to wildflowers or the seed mixes are sown in strips along the edges of fields; farmers are encouraged to use seed mixes which provide pollen and nectar sources over a long season. 'Many of the farms have entered an agri-environment scheme so get paid for the land they put over for wildlife although several farmers provide additional measures beyond such schemes. While some of the habitats required to meet the grade may already have existed on the farm prior to it being accredited, providing they meet the standards, that can count towards the 10% requirement,' says Shelley Abbott of Conservation Grade.

Conservation Grade doesn't require land to be converted to organic (though some members are organic), and the farms are free to use most artificial fertilisers and pesticides on their crops. 'Fair to Nature is about the creation and management of targeted habitats. On our non-organic farms, artificial fertilisers and pesticides are permitted with the exception of organophosphates for crop production and grain store treatment,' comments Ms Abbott.

The Red Tractor scheme (with a red tractor in an 'Assured Food Standards' double circle on products), the largest UK farm assurance body, uses independent assessors to check that food and drink meets their standards 'from farm to table'. Farms are inspected annually. The standards cover traceability – where the food has been farmed, processed and packed; food safety and hygiene; animal welfare; and environmental protection on the farm to minimise pollution and have the least impact on wildlife. It does not provide any standards for the protection or creation of wildlife habitats and it includes non-organic and organic farms.

With Conservation Grade, Red Tractor, LEAF, and organic having logos on products, all with different standards, the consumer is somehow expected to cope, and understand what all this means.

But is organic farming better for wildlife than non-organic farming? How do (usually) non-organic yet accredited wildlife-friendly forms of farming promoted by Conservation Grade and LEAF compare? Not that organic farmers necessarily retain more existing wildlife habitat on their farms or manage some of their land to give primary consideration to wildlife rather than to crop or livestock output; organic farmers are just as keen on agricultural production as non-organic farmers. However, a major study confirmed the findings of several others and showed that organic lowland farms are naturally richer in wildlife than non-organic farms.[xiii] The most obvious difference was for plants: species richness and the amount of natural plant cover on organic farms were both much higher, in the case of species numbers by more than 80% on organic fields. More diverse crop rotations and the absence of herbicides contribute to greater plant diversity.

Organic farms had more hedgerows per land area and those hedges were taller, wider, and had fewer gaps. The farms were also much more likely to have mixed livestock and arable (because the manure is needed as fertiliser for the crops) and that alone tends to increase their wildlife value simply by providing more of a mix of habitats. Using manure boosts invertebrate numbers and the bats and birds that feed on them. Spiders and ground beetles are more common on organic cropped fields purely because these fields are not treated with synthetic insecticides and herbicides that kill off any non-crop plants. The study found that both groups of invertebrates were also much more abundant in cereal crops sown in spring than in autumn, emphasising again the wildlife value (both for invertebrates and for birds that predate them) of spring-sown cereals. On organic farms, spring-sown cereals are usually the norm.

Butterfly abundance, and the number of species, is consistently higher on organic farms, both within cropped fields and in non-cropped marginal vegetation.[x] There is evidence from a range of studies that small mammals, bats, and farmland birds are more abundant in numbers and diversity of species on lowland organic farms.[iv] However, the research isn't always clear whether it is the variations in how the land and crops are managed, or the amount/quality of uncropped habitat that accounts for the difference. That isn't just an esoteric issue; it's important to know whether any practices taken for granted on organic farms can be transferred to non-organic farms to help boost their wildlife value too. Measures such as more care over hedge maintenance, leaving untreated wild plant margins alongside crops, and encouraging more of a mix of livestock farming with crop growing can all help improve the wildlife content of non-organic farms, as other chapters have illustrated.

A study of the wildlife value of a set of Conservation Grade farms, a similar set of established organic farms, and farms which had entered Entry Level Stewardship, a very basic but since discontinued wildlife management scheme (part of England's Environmental Stewardship scheme, now Countryside Stewardship), revealed significant gains for wildlife at both the Conservation Grade and organic farms.[xiv] Both Conservation Grade and organic farms supported 20%

more habitat diversity compared to Entry Level Stewardship farms and provided greater richness of butterflies and plants overall.

Butterfly species richness was significantly higher on organic farms (50% higher) and marginally higher on Conservation Grade farms (20% higher) compared with farms in Entry Level Stewardship. Organic farms supported significantly more plant species than Entry Level Stewardship farms (70% higher) but Conservation Grade farms were only 10% richer in plant species. There were no significant differences between the three schemes for species richness of bumblebees, solitary bees, or birds, possibly because all three groups are much more mobile and therefore use the landscape at larger scales than that of individual farms. As a consequence the authors suggested that perhaps if Conservation Grade or organic schemes were concentrated more in particular locations, rather than in a scatter of individual farms, positive effects for bumblebees and birds would be found.

The decision by individual farmers on whether or not they should seek registration as organic is primarily an economic one, though many farmers doing so – and those adopting other alternatives such as the LEAF standard – have concerns about the long-term sustainability of non-organic farming. There are worries in particular over intensive lowland farming which is so dependent on artificial inputs of pesticides and fertilisers to maintain high crop yields. Many too are anxious about the impact of their farming practices on wildlife, both directly on their land and indirectly in relation to issues such as fertiliser pollution of watercourses. Derided for years by some non-organic farmers as an oddity, organic farming is starting to provide some answers to seemingly intractable problems on non-organic farms, as fears mount over issues such as herbicide-resistant weeds, insufficient soil organic matter, and antibiotic resistance in livestock disease.

'Organic farming has a lot to offer because we have been having to do without agro-chemicals for the past 60 or 70 years,' said Helen Browning, the Soil Association's Chief Executive, at the South of England Farming Conference in November 2016. 'We have learned a lot of techniques that can help as the chemistry stops working.' She cited examples including addressing the problems of aggressive weeds like Blackgrass and crop pests such as Leatherjackets without using chemicals and how to manage no-till crop systems without using glyphosate to kill off the previous crop before drilling in the next one.

Organic farming, in order to attract premium prices for its produce, will continue to be a small component of UK farming, but it does provide undoubted and substantial benefits for farm wildlife. The UK and devolved governments need to promote and encourage more organic farming by focussing increasing support on the supply chain from farmers, via processors and wholesalers, to retailers. The financial supports aiding conversion to organic farming in all UK countries are substantial and these need to be retained once we leave the CAP. They should also be independently reviewed to examine whether they provide value for money and need modification to get more farmers in different farming sectors to convert to organic production.

Alternatives to becoming organic, such as LEAF and Conservation Grade, also promote wildlife-boosting practices; many more farmers should be urged to consider these schemes too. However, there is concern, that the UK's governments, Defra in England especially, are reluctant to be too effusive in promoting organic produce from UK farms – for example by encouraging its use in all schools – because of likely criticism from non-organic farmers and, in particular, from farming unions. It is high time that these attitudes are discarded and that all forms of more sustainable and more wildlife-friendly agriculture get the attention and focus that they clearly deserve.

Endnotes

i   http://www.organicresearchcentre.com/?go=Organic%20Research%20Centre&page=Organic%farming.

ii   'Organic farm incomes up in 2015.', Organic Research Centre press release 13 April 2016.

iii   'Lazy Man of Europe,' The Soil Association, 2011.

iv   'Organic Europe Country Reports,' Research Institute of Organic Agriculture.

v   'Organic Farming Statistics, 2015,' Defra, 2016.

vi   'Organic food sales surge to hit almost £2 billion,' *Daily Telegraph*, 23 February 2016.

vii   'Glastir Organic: How to Complete, Guidance 2017,' Welsh Government Rural Communities – Rural Development Programme for Wales 2014 – 2020. Welsh Government.

viii https://www.ruralpayments.org/publicsite/...schemes/agri-environment-climate-scheme.

ix   https://www.daera-ni.gov.uk/articles/agri-environment-schemes.

x   Environment, Food and Rural Affairs Committee, House of Commons: Written evidence from the Soil Association, 2013.

xi   http://www.leafuk.org/leaf/home.eb.

xii   http://www.conservationgrade.org/.

xiii 'Does organic farming affect biodiversity?' by Ruth Feber et al. *Wildlife Conservation on Farmland, Volume 1*. Edited by David Macdonald and Ruth Feber. Oxford University Press, 2015.

xiv 'Supporting local diversity of habitats and species on farmland: a comparison of three wildlife-friendly schemes,' Chloe J. Hardman et al, 2015. *Journal of Applied Ecology*, 53: 171–180. doi:10.1111/1365-2664.12557.

CHAPTER 11

BROCK AND THE SCOURGE OF **TB**

The Mole had long wanted to make the acquaintance of the Badger. He seemed, by all accounts, to be such an important personage and, though rarely visible, to make his unseen influence felt by everybody about the place.

KENNETH GRAHAME (1859–1932), *THE WIND IN THE WILLOWS* (METHUEN, 1908)

On a warm spring day full of sunshine, the rolling Pembrokeshire countryside where Stephen James, dairy farmer and current President of the National Farmers' Union (NFU) Cymru, lives and works is deceptively peaceful. But this part of South West Wales and much of South West England are the epicentres of one of the most intractable issues facing both dairy and beef farming in Britain. For these are the places where bovine tuberculosis (TB) is most prevalent and where the debate about the role badgers play in its transmission to cattle is the most acute.

'We never had TB in our cattle herd until 1993,' says Mr James, a quietly spoken and thoughtful man who, with his wife Joyce and son Daniel, farms 500 acres with a milking herd of 330 Holstein Friesians. 'Then a routine tuberculin test showed up positive for the disease. It was only one calf that had never been out in the fields since it was born; it was kept indoors during winter as all our cattle are but in a building badgers were then able to access. We have quite a few badgers on the farm; their setts are in the broadleaved woodland we have. So we had to have a few of our cattle culled to control the disease. It came as a bit of a shock.

'And since that first outbreak we've had TB in some of our cattle off and on over the years. The worst outbreak, though, was in 2012. But it was odd because the herd was tested in the May of that year and given the all-clear. In the October we wanted to sell 13 animals and they were pre-movement tested, part of the regulations in force before we are allowed to take them to market; we were

The source of a pinta. Stephen James on   his dairy farm in Pembrokeshire (courtesy of NFU Cymru).

expecting no sign of TB so we were stunned that six of them reacted positive to the tuberculin test. These cows were from a group of 45 cattle that had been grazing on one part of the farm and I decided to have them all culled in case the disease might spread to the rest of our herd. I couldn't risk that. In the end, 22 of the 45 were found to have TB lesions (affected organs which contain cheese-like nodules called tubercles) when they were examined after slaughter. So some were culled needlessly but it was the safer option. We were compensated and I don't have any grumbles about that.

'What we don't do now is to let any of our herd graze pasture near one particular badger sett because we reckon the badgers there have TB. And we are careful to make sure badgers can't get into the cattle sheds. Since 2012 we haven't had another outbreak. It's something you learn to live with; it bothered me a lot in 1993 when it happened first but I don't worry about it so much now. It's a great relief to have been free of it for a few years though. And, of course, I hope it stays that way.'

To most farmers the position is crystal clear: the seemingly inexorable rise in badger numbers over the last couple of decades – mammals that often feed on the very same pastures as cattle – explains why TB in cattle has become so commonplace. They don't want to see badgers exterminated but they do want to see their numbers greatly reduced. Many conservationists, on the other hand, argue that the majority of cattle get infected from each other, and that if badgers do play a role, it's not the major one. Along with animal rights groups, most conservationists aren't prepared to accept any culling of what is, after all, a fully protected species. Add to this the fact that the badger is the symbol of The Wildlife Trusts, and that the Trusts have no fewer than 800,000 members – almost twice the number of people working in farming – and you have a heady mix indeed. Not unexpectedly, passions run high.

Badgers certainly do cause farmers some headaches. Their setts aren't always dug in the most convenient places on farms. They also create latrines, distant from their setts, filled with faeces, mainly the digested remains of earthworms. Their digging for worms and ground-nesting wasps and bees can damage crops, and they often disturb poultry and gamebirds on their night-time farm walkabouts.

SO BLACK AND WHITE?

Eurasian Badgers, one of many species throughout the world, sleep during the day in one or more setts in their territorial range. The setts, which may house several badger families, have extensive systems of underground passages and chambers and have multiple entrances. Some have been in use for decades. It's particularly difficult to get an accurate idea of their numbers; the most recent estimate, for 2011–13, is of 72,000 social groups in England and Wales; at six per sett on average, that means 423,000 badgers (it could be more) and their numbers appear to have been increasing at about 2.6% per annum.[i]

Some of the highest densities in the UK are on farmland in southern England where they can exceed 15 per square kilometre; the UK average is a fraction of this. The lowland landscape created by modern farming – copses of broadleaved trees in which to dig setts and worm-rich pastures for feeding – is badger nirvana.

Badgers are very fussy over the cleanliness of their burrows, carrying in fresh bedding and removing soiled material; they defecate in latrines distant from the sett. If a badger dies within the sett, the others will seal off the chamber and dig a new one. Some badgers will even drag their dead out of the sett and bury them outside. Largely carnivorous, they feed mainly on earthworms but will eat a wide variety of plants and animals including large insects, small mammals, carrion, cereals, and root tubers. Litters of up to five cubs are produced in spring.

In general, badgers show considerable tolerance of each other both within and between groups. Fundamentally peaceful, badgers have been known to share their sett with other species such as rabbits and foxes, although they can be ferocious when provoked, a trait which has been exploited in the now illegal blood sport of badger-baiting.

Badgers are fully protected in the UK and so are their setts. Culling – or disturbance of a sett – can only be done if licensed by the appropriate wildlife protection organisation: Natural England, Natural Resources Wales, the Northern Ireland Environment Agency, or Scottish Natural Heritage.

A culprit? European Badger out in cattle-grazed pasture (courtesy of Kallerna).

For decades, control of TB in cattle has been predicated on regular testing using the tuberculin skin test. The aim is to trigger and then measure an allergic reaction to proteins and other antigens extracted from the bacteria that cause TB when they are injected into the deep layers of the animal's skin. Those that show a skin reaction are deemed to have the disease and are removed for slaughter. But the test is notoriously inconclusive. In Wales both positive and inconclusive reactors are immediately sent for slaughter; cattle in the herd not reacting are then isolated and re-tested later. Cattle in England with an inconclusive test result are left in the herd to be re-tested later. Research has shown that the test misses perhaps one in five TB-infected cattle,[ii] which supports the measures adopted in Wales. Moreover, although a negative TB test is compulsory before cattle can be moved off a farm (to a sale for instance), an imperfect test is bound to result in some infected cattle being moved, thereby potentially infecting others if they are bought by another farmer rather than sold for slaughter. Farmers receive government compensation for any cattle prematurely slaughtered.

Carcasses of culled cattle that show positive in the tuberculin test are inspected for TB lesions, and samples are sent to confirm whether or not the animal had bovine TB. If the disease is restricted to small areas of the carcass, those parts are removed and the rest can be processed normally for eating. If extensive infection is found, the whole animal carcass is disposed of and doesn't enter the food chain. Regular cattle testing has ensured that advanced cases of bovine TB are rare. Supermarkets have refused to accept any meat from TB-infected cattle, but it is apparently sold to some caterers and food processors and is used in a range of food products including pies and pasties. Eating beef from cattle that have had bovine TB is perfectly safe; the chance of contracting the disease from cooked meat is negligible.

The incidence of bovine TB has been increasing since about 1980, especially in South West England and South West Wales. The number of cattle slaughtered in Britain because they tested positive for TB was 39,915 in 2016 out of a total cattle herd of nearly ten million; three-quarters of those slaughtered prematurely were in England (mostly in the South West and North), almost all the remainder were in Wales.[iii] Over the last decade, an estimated £500 million has been spent in the UK on TB testing, compensation, and research.

Scotland, meanwhile, was designated by the EU as bovine TB-free in 2009 and remains so. There the incidence and risk of bovine TB has historically been very low and there is no evidence of a wildlife reservoir of the disease. The Scottish Government has, in recent years, introduced a stringent package of measures to retain this TB-free status.

There are probably many reasons why the prevalence of TB in cattle has increased in the rest of the UK. Cattle movements were not restricted until more recent years, thereby allowing transmission from one herd to another and potentially from one end of the UK to the other. Today's larger herds don't help either; once one animal picks up an infection, it can easily spread to many more. During the foot-and-mouth disease outbreak in 2001, when ten million sheep and cattle were slaughtered because the disease was highly contagious, TB testing was suspended. This was done to restrict access into the countryside to help reduce the spread of foot-and-mouth and many culled cattle herds were restocked without knowledge of the bovine TB status of the replacement animals.

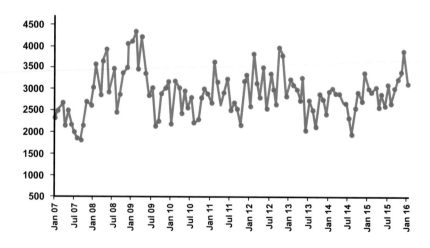

Chart showing number of cattle slaughtered for TB in UK per month. (Source: Defra.)

With all the media attention focussed on bovine TB and the role of badgers, it isn't surprising that most people assume that the disease is *the* major cattle disease in Britain. Not so. Monitoring data collated by a leading UK farming consultancy shows that, for dairy cattle culled prematurely, large numbers are culled because of mastitis (a potentially fatal udder infection), many others because they have failed to produce a calf (thereby limiting their milk output), while accidents and trauma cause some early deaths too.[iv] Cattle suffer from many other diseases: viral diarrhoea for example, and Johne's disease, which is a contagious, sometimes fatal, infection that primarily affects the small intestine.

Bovine TB also isn't new. It was a major problem in the UK in the 1930s when 40% of dairy cattle UK-wide were infected and there were 50,000 new cases of the human *M. bovis* infection every year, mostly transmitted via unpasteurised milk.[v] The disease was often fatal. The first scheme of testing and slaughter (with compensation) of infected animals to bring it under control was introduced in 1935. By 1979, bovine TB reached its nadir; just 0.02% of cattle tested positive.

So where do badgers fit in this deadly picture? TB was first discovered in badgers in 1971; maybe they caught it originally from cattle. The disease can move between the two species very easily, particularly when they feed in the same pastures. Deer, often grazing on pasture at night, might act as vectors too, although badgers are far more abundant than deer in the countryside and the incidence of TB in deer is very low; for example, only about 3% of deer in Wales are known to carry TB.[vi]

Badger habitat perfection. Woods and copses for sett building;
pastures for earthworms (Widecombe-in-the-Moor, Devon).

THE TROUBLESOME BACTERIUM

Mycobacterium bovis is a slow-growing bacterium and the cause of tuberculosis in cattle (bovine TB). Related to *M. tuberculosis* – the bacterium which causes tuberculosis in humans – *M. bovis* can also infect species other than cattle: humans, deer, llamas, pigs, domestic cats, foxes, badgers, goats, and several rodents. It is hard to spot bovine TB in its early stages as the signs are similar to other diseases and normally only develop in advanced stages of infection. It's usually picked up in the compulsory cattle testing programme before clinical signs appear, though early signs include cattle getting thinner, a recurring light fever, and having a reduced appetite.

The disease can be spread in exhaled air, sputum, urine, faeces, and pus. It can be transmitted by direct contact; via sputum or excreta on a pasture or feeding trough; or by inhalation of aerosols. Once a cow is infected, however, it does not immediately start spreading the disease; TB develops very slowly and it takes time for lesions to grow in the lungs, and these lesions have to open up before cattle start coughing out the bacteria.

Luckily, the bacterium is readily killed when milk is pasteurised.

By no means all badgers have TB and, in those that do, it isn't a major cause of death. TB-infected badgers can live for several years during which time their ability to infect other animals will vary. The incidence of the disease in badgers also ranges enormously from almost zero up to a maximum of 38% of animals, higher in males than females, and there is evidence (based on the analysis of different TB bacterial strains) showing that cattle to cattle transmissions are more prevalent than badger to cattle transmissions.[vii] In Wales, less than 7% of badgers found dead since September 2014 tested positive for TB.[vi] Another study found that transmission from badgers to cattle might account for no more than 6% of cattle outbreaks.[viii] The overwhelming majority of cattle infections were passed on by other infected cattle, either directly or indirectly, although the links between the density of badger populations, the prevalence of TB amongst them, and the rates of TB in cattle are still not fully understood.

Nevertheless, in areas of England where TB in cattle is prevalent, badger culling has been a common response by successive UK governments (there has been no badger culling in Wales though that might change in 2018), its critics claiming that it has been carried out largely to satisfy the demands of farmers and show that at least 'something is being done'. Its proponents argue that it will help significantly to reduce the occurrence of the disease in cattle. So far, with a substantial amount of badger culling at different times in several areas of England, there is no evidence that it has been very effective at all; TB rates in cattle in England remain high and are in fact increasing.

There is widespread public support for an alternative to culling. In October 2012, MPs voted 147 in favour of a motion to stop a 2012/2013 badger cull with 28 voting against. The debate had been prompted by a petition on the government's e-petition website; by the time it closed on 7 September 2013 it had half a million signatures.

Only one scientifically-based and thoroughly-monitored badger culling trial has ever been set up to see whether killing badgers has an impact on the incidence of bovine TB in cattle. It was rather clumsily entitled the Randomised Badger Culling Trial (RBCT). Designed and overseen by an Independent Scientific Group (ISG), this £50 million trial funded by the UK's then Labour Government ran from 1998 to 2007 and took place in ten groups of three areas in South West England where bovine TB was particularly prevalent. Each group was 100 square kilometres in size. One area in each group had proactive killing in which as many badgers as could be found were culled to get their numbers as low as possible. In the second area in each group, badgers were culled on and around farms at which TB outbreaks in cattle had occurred (reactive culling). In total, 11,000 badgers were killed with a shotgun at point blank range after they had been caught in baited cage traps. To act as a control, the third area had no badger culling.

Analysis of the information from the trial showed that the frequency of TB in cattle herds in the proactive badger-culling areas was reduced by up to 40% compared with that in the control areas where no badger culling took place.[ix] However, that reduction was short-lived, presumably because badgers re-established in the areas where they had been killed. By 2011, four years after culling stopped, the incidence of cattle TB was just 10% lower in the proactive badger-culling areas.[x] To achieve any longer term benefit for cattle, badger culling would need to be repeated at regular intervals.

In a zone up to two kilometres outside and around the proactive culling areas, the incidence of cattle herds contracting TB was nearly 8% higher than in the controls. This unexpected result came from the disturbance of the remaining badgers, which changed their behaviour and distribution in unpredictable ways, increasing the spread of TB to cattle rather than decreasing it. The effect, though, was relatively short-lived; it had dissipated within two years.[xi]

The reactive badger culling – around farms where cattle had TB – had to be stopped mid-trial because the incidence of cattle TB in these areas actually *increased* by 22%, an effect again seemingly caused by badgers spreading the disease through moving in unpredictable ways. It was the opposite of what farmers were hoping to achieve and left Defra officials and the Government dumfounded.

An independent review of the RBCT concluded that 'the financial costs of culling an idealized 150 square kilometre area would exceed the savings achieved through reduced cattle TB by a factor of up to 3.5 times'.[xi] Another analysis of figures published by Defra found that for every square kilometre of land, culling badgers had cost nearly £13,000 to save an estimated £714 in reduced cattle slaughter compensation and TB testing costs.[xii] So such large-scale badger culling was an economic non-starter. The ISG concluded that 'badger culling cannot meaningfully contribute to the future control of cattle TB in Britain' and that 'some policies under consideration are likely to make matters worse rather than better'.[ix] They also pointed out that cattle to cattle transfer was the main cause of disease being spread to new areas, not badgers.

Nevertheless, in 2012, the Conservative-Liberal Democrat Government decided on two further badger culls, one in Somerset and one in Gloucestershire in 2013. This was based on advice from a panel of leading experts convened by Professor Sir Bob Watson, then Chief Scientific Advisor at Defra, and Nigel Gibbens, the Chief Vet. Their view was that if the badger cull area was large enough (over 150 square kilometres) the perturbation effect would be minimised – the bigger the area, the smaller, as a proportion, the perimeter becomes – and an estimated reduction in cattle TB incidence of 16% could be achieved over a nine year period. Strangely, the group made no mention of any need to repeat the badger culling as the benefit in bovine TB reduction would reduce after several years because other badgers would refill vacant territories. To achieve even this relatively small decrease in TB, a high proportion of the badgers present in any culling area would have to be killed, probably over 70%, and the costs, including policing costs to deter the inevitable protestors, were bound to be high.

At more or less the same time – and reviewing precisely the same evidence pointing to its limited effectiveness – the Welsh Government decided not to go ahead with any badger culling! Controversially, the new culls in England were designed specifically to test night shooting of badgers (paid for by landowners, not all of whom wished to participate). Much of it was done by landowners and tenants, something animal rights groups were strongly opposed to because of the predicted high risks of injury and lingering badger deaths. Substantial public opposition disrupted the culls and caused the costs to soar. Little over a year after they started, the culls were scaled back.

An expert independent panel chaired by Professor Ranald Munro, the former Head of Pathology at the Veterinary Laboratories Agency, reported in 2014 to the Secretary of State for Environment, Food and Rural Affairs (at that time Owen Paterson) on the effectiveness of the cull. It concluded:

> Controlled shooting alone (or in combination with cage trapping) did not deliver the level of culling set by Government. Shooting accuracy varied amongst contractors and resulted in a number of badgers taking longer than five minutes to die, others being hit but not retrieved, and some possibly being missed altogether. In the context of the pilot culls, we consider that the total number of these events should be less than five per cent of the badgers at which shots were taken. We are confident that this was not achieved.[xiii]

Defra ignored these findings, scrapped the panel, and continued the badger cull for another two years!

The latest plan for bovine TB control in England, produced by Defra, for the period 2010 to 2015, and updated since, involves annual testing of all cattle in defined high-risk areas (mainly South West England, the west of England, and East Sussex) and areas adjacent to them, with four year testing in low-risk areas (the rest of England).[xiv] Cattle can't be moved off farms in the high risk and adjacent areas unless they are shown by testing to be TB free.

Yet badger culling still hasn't been abandoned in England in spite of its limited effectiveness, substantial costs, and concerns that some of the culling is cruel. In August 2015, the UK Government announced that culling would begin anew in Dorset and continued in Gloucestershire and Somerset. Once again it was opposed strongly by organisations such as the RSPCA and animal rights groups, some of whose members attempted to interfere with the shooting. No improvements to humaneness were reported following this second year of culling, and the British Veterinary Association withdrew their support. The culls had been both ineffective and inhumane. Nevertheless, culls using licensed marksmen continued or were initiated in 2016 in parts of Cornwall, Herefordshire, Devon, Gloucestershire, Somerset, and Dorset and more are planned in 2017 and 2018 for these counties and others. Since 2013, around 15,000 badgers are known to have been culled in England.[xv]

To date there has been no reported decline in cattle TB rates and no monitoring of the impact of better biosecurity on farms within the cull zones to know which aspects of England's TB control policy are achieving any success. Nevertheless, after the 2015 culling, Nigel Gibbens saw fit to release a statement of advice that, 'The outcome of this year's culls indicates that industry-led culling can deliver the level of effectiveness required to be confident of achieving disease control benefits'.[xvi] It's difficult to understand on what evidence such a statement was based.

'It's clear that testing cattle frequently is the most effective way of reducing bovine TB,' comments Professor Matthew Evans of Queen Mary University of London. 'In areas where testing is more frequent, TB is beginning to decline. Farmers and policymakers should not ignore this evidence which is based on the Government's own data. You could cull all the badgers and you would still have TB simply due to the fact that cattle give TB to one another and the TB test is imperfect. The disease will continue to spread because we move cattle around while badgers tend to move little unless disturbed. Badger culling as practised at present in England is not scientific, it is not being monitored, and is unlikely to have a positive effect. Badgers are not without culpability in TB; they will act as foci of infection locally, they can maintain infection and will re-infect cattle. But they will not spread the disease widely.'

Using state-of-the-art computer modelling to understand how the interaction of different factors impacted on cattle infection rates, Professor Evans and colleagues found that it is the frequency with which cattle are tested for TB and whether or not farms utilise winter housing for livestock that have the most significant effect on the number of infected cattle.[xvii] 'Investing in more frequent cattle testing – as is being done in Wales – is a far more effective strategy than badger culling. Our analysis showed that TB in cattle is rising in England but falling in Wales; the Welsh policy was likely to lead to a reduction in TB but the English policy was not,' he said.

The economic costs of bovine TB are shared by farmers and governments. Costs are incurred because early cattle slaughter reduces the value of the animal. The government pays farmers compensation for slaughtered animals based on their market value; farmers with insured livestock can claim less compensation. Although vets' fees for carrying out tests on the herd are paid for by the government, other costs are incurred by farmers. When an animal in a herd tests positive for the disease, the whole herd is put under movement restrictions until all the remaining animals are certain to be negative. Farmers are then unable to move their cattle to market or buy in replacements for animals that are slaughtered.

In Wales there hasn't been any programme of badger culling. The Welsh Government has acknowledged that the main route of transmission of TB is from cattle to cattle. All cattle have therefore been tested annually for TB since 2008; pre-movement testing for bovine TB has been mandatory in Wales since 2010. There is a much more vigorous policy of quickly removing cattle off farms for slaughter if their tuberculin test is positive or inconclusive. The amount of testing has been increasing year on year: in 2002 there were about 700,000 cattle tests done in Wales; in 2014 it was 2.15 million.[vi] Other TB sources are checked too: domesticated animals such as llamas and alpacas, and wild animals, including badgers and deer, that are killed on roads. Cattle farmers not implementing advice from vets or not sticking to the testing regulations have their compensation reduced if any cattle have to be slaughtered. It's a more rigorous approach than that being adopted in England.

So far, it seems to be working. Welsh Government data shows that 95% of cattle herds in Wales were TB free in 2017 which means that at least 200 fewer herds were under movement restrictions because of a TB incident compared to three years ago.[xviii] And the number of new incidents has fallen by over 40% since its peak in 2009. It's a record far better than that in England where the number of new incidents continues to rise.[xix] It is currently higher than it has ever been!

Modified plans were announced by the Welsh Government in the summer of 2017.[xx] From October 2017, Wales will be divided into low incidence, intermediate, and high incidence TB zones. Post-movement testing will be done for the first time in the low and intermediate zones but persistent problem herds in high incidence zones will have individual action plans that will include humane, local badger culling if testing shows the badger population to be infected. Badgers testing negative will be microchipped and released and there will be no large-scale badger culling as in England. Whole herd tests at six-monthly intervals for herds with a long history of TB will continue; there will be more restrictions on herd movements and cattle reacting inconclusively to the tuberculin test will still be culled.

SECURITY, SECURITY

Many of the solutions to the bovine TB problem lie with cattle farmers themselves. There are a vast number of common sense precautions that an individual farmer can put in place to reduce the chances of his cattle getting infected, so-called 'biosecurity'. Cattle need to be kept away from those on adjacent farms so fences between farms must be suitably stock-proof; double boundary fences a few metres apart should be considered to prevent nose-to-nose contact on shared boundaries. Wherever possible, farmers need to prevent cattle access to shared watercourses such as ponds or streams and provide piped water to drinking troughs instead. They also need to know where any cattle they buy at sales have come from and get advice about animal health and previous testing before purchasing them. These are all sensible and practical measures the NFU supports.

The Welsh Government has given grants to livestock markets to enable them to set up displays to show the disease and testing history of an animal and the herd it comes from so farmers know more about the status of stock they buy. Buyers also need to isolate incoming cattle, for instance in a field where no contact can be made with other cattle already on the farm. There are other ways to reduce TB too: keeping indoor housing for cattle well ventilated and not cramming too many cows together. They should be kept away from freshly spread cattle manure or slurry (the TB bacterium can survive in stored slurry for six months) and farmers need to dispose of cattle bedding so that the animals can't get access to it. Farm machinery needs to be thoroughly cleaned and disinfected, particularly if sharing it with a neighbouring farm, and farmers need to insist that contractors do the same. How much of this advice is being taken on board by farmers isn't clear; there isn't any monitoring of its effectiveness and none of these practical measures are mandatory.

The Welsh Government has also established a large 'Intensive Action Area' (IAA) located in north Pembrokeshire and other parts of South West Wales where the disease is particularly prevalent among both beef and dairy cattle. Since 2010, all cattle herds in the IAA have been TB tested twice a year and by 2017, the incidence of cattle TB had fallen by 35%.[xxi] Badgers throughout the IAA are being trapped and vaccinated with the BCG vaccine, the one given to protect children from TB. So far, well over 5,000 doses of vaccine have been administered over four years, although this part of the programme is on hold because of a worldwide shortage of the BCG vaccine.

Vulnerable. Beef cattle in Devon, one of the most badly affected bovine TB counties.

Research commissioned to assess the implications of this holdup has concluded that even without completing a fifth year, the four years' worth of effort already undertaken has been beneficial in lowering badger TB prevalence in the IAA.[xxii] The model simulation showed that the greatest reduction is achieved in the first few years of the vaccination; after that the absolute gains of each subsequent year of vaccination decrease. No disease has been found in any goats, llamas, or alpacas in the IAA and the incidence of TB in badgers killed on roads has fallen since vaccination began. But the road death numbers are very small (around 30 per year) and it's too early yet to tell if this fall will continue.

Vaccinating badgers is an attractive option; it's unlikely to require policing because it has wide public support, a good number of people are likely to volunteer their time (after training) to help do it, and several voluntary wildlife conservation organisations have been doing it in their areas. It's a far greater 'win/win' opportunity than badger culling will ever be. Although vaccination doesn't prevent TB infection, it reduces the progression of the disease substantially, decreasing its severity and lessening the excretion of TB bacteria which cattle can pick up on pastures or elsewhere. A vaccination scheme needs to be implemented annually to protect new cubs but no one knows for how many years it would need to be maintained before the disease was eliminated. Using an oral vaccine applied to bait and left at badger sett entrances might sound a more attractive – and a very much cheaper – option than trapping badgers to inject them with the vaccine. In reality though, it would be impossible to ensure that badgers took the right dose and that enough of them were protected.

'The cost of vaccination could be reduced by providing grants to farmers and landowners to carry out some of the labour themselves,' said Professor Christianne Glossop, Chief Veterinary Officer for Wales in 2015.[xxiii] 'TB is expensive. This year we're spending £25 million in Wales on cattle testing, on compensation to farmers, on breakdown management, and on a new programme we're rolling out to get private vets more involved in supporting their clients. So against that backdrop the notion that vaccinating badgers is expensive, it's just one piece of the whole programme. The most expensive part of vaccinating is labour: walking fields, finding the badgers, and catching them to vaccinate them.'

The Welsh strategy shows that high-intensity testing on an annual basis, or six monthly where necessary; movement restrictions; removing cattle with positive or inconclusive results quickly and effectively; and ensuring advice and action on farms is followed, can result in the identification and removal of the reservoir of infected cattle. Then there is much less chance of onward transmission.

Nevertheless, NFU Cymru doesn't agree that the Welsh strategy is broad enough and it remains committed to the view that badger culling is necessary. They have welcomed the recent inclusion of very limited local culling if it proves necessary. 'Bovine TB continues to cause untold heartache and stress to cattle farmers across Wales, and places an enormous emotional and financial strain on farming families,' says Peter Howells, their Farm Policy Advisor. 'This is a complex disease that must be tackled in the round, including addressing wildlife disease reservoirs, if we are to stand any chance of eradicating it. Farmers are playing their part in tackling bovine TB through cattle-based measures, but the reservoir of infection that exists in wildlife has not been confronted. The disease must be actively removed from the badger population in areas where bovine TB is endemic.

'From its inception, NFU Cymru has consistently raised concerns about the cost and effectiveness of the Welsh Government's badger vaccination policy in the IAA in North Pembrokeshire. It is NFU Cymru's view that a bovine TB wildlife strategy predicated solely on the vaccination of badgers is not a viable or sustainable policy. We agree that cattle movement controls, cattle testing, vaccination if available, and biosecurity all have a vital role to play in a TB eradication plan, but experience from across the globe has shown that a genuine TB eradication plan must also include a strategy for dealing with the disease reservoir in wildlife in areas where it is endemic.'

So why vaccinate badgers but not vaccinate the cattle themselves to protect them against the disease? The problem is that vaccinating cattle against TB is not permitted within the EU because the vaccine interferes with the skin test, making it almost impossible to tell whether a cow is genuinely infected with the disease or has been protected against it with the vaccine and has acquired antibodies that way. Research is underway to devise a way of showing up this difference – a so-called

DIVA test – but it is expensive and not yet widely available. Either way, there is no provision under the relevant EU Directive for certifying meat or meat products as being from 'TB-free but vaccinated herds' so the UK would be banned from selling beef, dairy products, and live cattle to any EU country, a substantial market. When we leave the EU, our sales to EU countries will remain under the same restriction. In any case, as with human disease vaccinations, cattle vaccination against bovine TB is not going to be completely effective; the efficacy of the BCG vaccine in cattle is only between 50% and 70%.[xxiv]

The UK Government, responsible for dealing with TB in England, is still fixated on badger culling. Liz Truss, the then Environment Secretary, told the NFU conference in February 2016 that the cull had to be expanded: 'I want to see culling expanded across a wider number of areas this year. The Chief Veterinary Officer's advice is that this is the only way to secure the full benefits of our comprehensive strategy,' she said.[xxv] The NFU supports it. 'The Government's 25-year TB eradication strategy is the first comprehensive plan of its kind we have ever had to tackle bovine TB in England and the NFU believes it gives us the best chance of controlling and eradicating this disease. It is based on the best available evidence and scientific advice and includes all the elements needed to give us the best chance of achieving a TB-free England – cattle testing; cattle movement controls; on-farm biosecurity; badger vaccination; cattle vaccination; and culling of badgers in areas where TB is rife. If nothing is done to deal with the disease in the wildlife reservoir at the same time as it is dealt with in cattle then re-infection will continue to occur, the disease will continue to spread, and herds that have been clear of TB will continue to be affected by it,' says an NFU spokesperson.

Ms Truss had presumably taken far more notice of the NFU than she had of a letter to The Guardian in September 2015 from a set of leading scientists with expertise in environmental issues, veterinary medicine, wildlife, and livestock health and welfare, saying that badger culling is still opposed by the majority of scientific experts and called on the Government to immediately reconsider its decision to continue with it.[xxvi] They went on to say that there is considerable research evidence and experience demonstrating the central importance of cattle-to-cattle transmission, both within and between herds, in maintaining and disseminating the disease.

The report summarising the impact of two years (2014–2015) of farmer-led badger culling in parts of Somerset and Gloucestershire concluded that it was impossible to assess whether it had reduced bovine TB in cattle in those areas, partly because there were no controls where badgers weren't killed to compare with, there was no monitoring, and most of the badger killing was haphazard.[xxvii] Consequently, there was no proven benefit, presumably a result that Ms Truss also took no notice of.

Some advocates of badger culling as a means of significantly reducing the incidence of the disease in cattle frequently point to the experience of badger culling in the Republic of Ireland (ROI). Undoubtedly, a cursory look at the numbers looks impressive; badger culling – much of it done inhumanely using snares – has been carried out since the mid 1980s, and the number of TB-infected cattle has fallen, from nearly 29,000 in 1997 to under 16,000 in 2013. But working out whether this fall is linked to badger culls is well nigh impossible because several other measures have also been introduced over the years.[xxviii]

The Irish Government claims only that badger control 'has contributed' to the drop in TB, and their stated intention is to replace badger culling with badger vaccination when that becomes a practical option. Irish scientists are leading the way on researching an oral vaccine. In Northern Ireland, cattle with TB have also declined even though no badgers have been culled there. The Irish experience hardly justifies the 'clear evidence' that badger culling is the way forward for TB control in cattle, as claimed in 2013 by Owen Paterson when he was Secretary of State for Environment, Food and Rural Affairs at Defra.[xxix]

Professor John Bourne, an eminent animal disease scientist who chaired the expert group that oversaw the RBCT, has been particularly outspoken. He has publicly accused Defra of continuing to either ignore, cherry pick, or purposefully misinterpret the science on badger culling. He argues that while cattle-control measures have been strengthened, they are still inadequate. 'Defra fails to fully appreciate that this is primarily an infectious disease of cattle and that the tuberculin test is very insensitive. As a consequence large numbers of infected cattle remain undiagnosed and perpetuate the disease in infected herds as well as spreading the disease to other cattle herds and wildlife,' he said in press interviews in September 2015.

In England, the onerous – and probably thankless – task of issuing licences to cull badgers falls to Natural England, the Government's conservation advisers in England. They operate this aspect of their work under legally binding 'guidance' provided to them in 2015 by Liz Truss when she was Secretary of State for Environment, Food and Rural Affairs.[xxx] In doing so, Natural England has numerous issues to consider, including what are termed 'reasonable' biosecurity measures being taken on participating farms; that the cull area exceeds 100 square kilometres (to minimise edge perturbation); that the cull is in Defra's mapped 'High Risk' or 'Edge Areas'; that 90% of the land is accessible for culling; that culling is sustained for four years; that it kills at least 70% of all badgers; and that any shooting of free ranging badgers is done by people who have completed a shooting training course approved by Government.

Dominic Dyer, CEO of the Badger Trust, was very direct in his criticism of England's continuing badger culling: 'The badger cull is built on three pillars of sand: incompetence, negligence, and deceit, and will ultimately collapse because it fails to address the key cause of bovine TB which is cattle to cattle infection. We could kill every badger in England but bovine TB will continue to spread in cattle herds due to inaccurate TB testing, excessive numbers of cattle movements, and poor biosecurity controls.'[xxxi]

NATURAL ENGLAND UNDER THE COSH?

In granting culling licences, Natural England doesn't appear to consider whether culling badgers is known to be effective as a means of reducing TB in cattle. Nor are they willing to provide any evidence showing what science they believe substantiates badger culling as an effective means of reducing the long-term incidences of bovine TB in cattle. 'Natural England's role as the licensing authority is to operate within government policy' is all that Natural England would divulge to me after I made a Freedom of Information request to try to find out! Their Chief Scientist, Dr Tim Hill, would not even communicate with me, a dire position for a public servant in a public body to adopt. In effect, Natural England, an organisation that allegedly prides itself on the veracity of its science to support its policies, is acting as the front for licencing badger culling based on an equally ill-informed Defra decision that culling will help control this cattle disease. It is Defra, of course, that funds all of Natural England's work!

So, what is the solution to eliminating bovine TB, or at least to reducing it to acceptable proportions? Long term, it's likely to rely on an improved targeted vaccine for cattle combined with a more sensitive test that can differentiate between animals that have contracted TB and those that have been vaccinated against it. Its development and testing is by no means imminent.

In the meantime, and as a vital part of any long-term strategy, overwhelming expert opinion suggests strongly that reducing cattle to cattle transmission by implementing much tighter biosecurity arrangements on all beef and dairy farms in all parts of the UK deemed to be at any risk of bovine TB is essential. Why, for instance, has Defra waited until 2016 to introduce statutory post-movement cattle testing for cattle entering England's Low Risk Areas to reduce the risk of importing TB-infected animals from higher risk areas? According to Neil Blake, President of the British Cattle Veterinary Association, 50% of new herd TB incidents in low risk areas are caused by purchased animals! Even Defra admits that farmers themselves aren't doing enough to reduce TB spread. In the English Midlands, the number of bovine TB cases increased from 80 in the first half of 2015 to 90 during the first half of 2016 as a result of inward cattle movements from other parts of the country, wildlife, and an 'undetermined source' in roughly equal proportions.[xxxii]

The evidence to support badger culling as an effective part of the strategy – other than achieving a temporary reduction in cattle disease for a few years but only if the vast majority of badgers are culled over large land areas – simply doesn't stack up. Maybe vaccination of badgers (when it resumes) will reduce cattle TB as well as the incidence of the disease in badgers; the early indications are promising. But badger vaccination is only likely to be cost effective if vaccine-laced baits can be developed and deployed efficiently, and that's not likely to be easy.

While the impact of premature cattle culling on individual farms can be substantial and highly emotive, it is worth remembering that farmers are compensated financially (albeit not always fully), and that most of the carcass meat enters the food chain as it would when the same cattle are eventually due for slaughter anyway, a fact that is rarely mentioned. It's time the UK Government admitted that their continued badger culling in parts of England where the disease in cattle is particularly rife is not organised to produce objective evidence of its efficacy, is unacceptable to a large proportion of the public, and, so far as can be ascertained from the one and only objective scientific study (the RBCT), contributes little or nothing to reduce cattle TB long term.

No one doubts that badgers infect cattle – and that cattle infect badgers and one another – but all the objective evidence suggests strongly that culling badgers is not going to play a significant role in eliminating cattle TB.

Endnotes

i   'Badgers in the rural landscape – conservation paragon or farmland pariah? Lessons from the Wytham Badger Project,' David Macdonald et al. *Wildlife Conservation on Farmland, Volume 2*. Edited by David Macdonald and Ruth Feber. Oxford University Press, 2015.

ii   www.cam.ac.uk/research/news/minimising-false-positives-key-to-vaccinating-against-bovine-tb#st hash.0VmKqvzh.dpuf.

iii 'Quarterly publication of national statistics on the incidence and prevalence of tuberculosis in cattle in Great Britain to end December 2015,' Defra, 2016.

iv   'How Healthy Are We Now?' Kite Consulting, 2013.

v   ' A review of tuberculosis science and policy in Great Britain,' Debby Reynolds, 2006. *Veterinary Micro biology*, 112(2–4):119–126.

vi   'Welsh Government TB Dashboard. Quarter 4, 2015,' Welsh Government, 2016.

vii 'Badgers and bovine tuberculosis: beyond perturbation to life cycle analysis,' by David Macdonald et al. *Wildlife Conservation on Farmland, Volume 2*. Edited by David Macdonald and Ruth Feber. Oxford University Press, 2015.

viii 'The Contribution of Badgers to Confirmed Tuberculosis in Cattle in High-Incidence Areas in England,' C. Donnelly and P. Nouvellet, 2013. PLOS Currents Outbreaks, Edition 1.doi:10.1371/cur rents.outbreaks.

ix   Bovine TB: A Science Base for a Sustainable Policy to Control TB in Cattle: An Epidemiological Inves tigation into Bovine Tuberculosis Final Report of the Independent Scientific Group on Cattle TB, 2007.

x   'Analysis of further data (to 28 August 2011) on the impacts on cattle TB incidence of repeated badger culling,' Christl Donnelly et al, 2011.

xi   'The duration of the effects of repeated widespread badger culling on cattle tuberculosis following the cessation of culling,' Helen Jenkins et al, 2010. *PLoS One*, 5(2): e9090.

xii 'Badger cull maths, stats and management,' Dr Tim Hounsome. www.badgergate.org/need-to...culls/badger-cull-maths-stats-management, 2014.

xiii 'Pilot Badger Culls in Somerset and Gloucestershire.' Report by the Independent Expert Panel, 2014

xiv The Strategy for achieving Officially Bovine Tuberculosis Free status for England. Defra, 2014. 2010 to 2015 Government Policy: Bovine Tuberculosis (bovine TB), Defra.

xv 'Groups in final preparations for badger cull,' *Farmers Weekly*, 1 September 2017, page 6.

xvi https://www.gov.uk/government/uploads/system/uploads/attachment_data/file/486612/cvo-advice-badger-culls-2015.pdf.

xvii 'Coupling models of cattle and farms with models of badgers for predicting the dynamics of bovine tuberculosis (TB),' Aristides Moustakas & Matthew Evans, 2015. *Stochastic Environmental Research and Risk Assessment, volume 29*, Issue 3: 623-635.

xviii 'Welsh Governments sets out new plans for bovine TB eradication but rules out England style badger cull,' *Wales Online*, 19 October 2016.

xix Quarterly publication of National Statistics on the incidence and prevalence of tuberculosis (TB) in Cattle in Great Britain – to end March 2017. Defra, 14 June, 2017.

xx http://www.walesonline.co.uk/business/farming/welsh-government-reveals-plans-tackling-13214054.

xxi 'New incidence of TB down 35% in IAA,' Welsh Government press release, 25 May 2017.

xxii 'Simulations of the effect of badger vaccination on bovine TB in badgers and cattle within the IAA, 2016,' Animal and Plant Health Agency.

xxiii http://www.bbc.co.uk/news/uk-wales-33622465.

xxiv 'Field evaluation of the efficacy of Mycobacterium bovis BCG against bovine tuberculosis in neonatal calves in Ethiopia,' G. Ameni et al, 2010. *Clinical and Vaccine Immunology*, 17 (10): 1533-1538.

xxv https://www.gov.uk/government/speeches/environment-minister-addresses-the-national-farmers-un ion-conference.

xxvi 'Badger cull is flawed and must stop now,' *The Guardian*, 2 September 2015.

xxvii Report of the incidence of bovine tuberculosis in cattle in 2014 - 2015 in the areas of Somerset and Gloucestershire exposed to two years of industry-led badger control. Animal and Plant Health Agency, 2016.

xxviii Channel 4 Fact Check: Has the Irish Badger Cull Worked? 7 March 2014.

xxix https://www.gov.uk/government/speeches/environment-secretary-owen-paterson-address es-the-2013-oxford-farming-conference.

xxx Guidance to Natural England: Licences to kill or take badgers for the purpose of preventing the spread of bovine TB under section 10(2)(a) of the Protection of Badgers Act 1992. Issued by Defra, 17 December 2015.

xxxi http://www.theecologist.org/News/news_round_up/2988047/englands_100m_badger_cull_extensions_condemned.html.

xxxii 'Mid-year (first six months) Descriptive Epidemiology Report: Bovine TB Epidemic in the England Edge Area,' Defra Animal and Plant Health Agency, 2016.

CHAPTER 12

WHERE THERE'S MUCK

The fairest thing in nature, a flower, still has its roots in earth and manure.

D. H. LAWRENCE (1885–1930), ENGLISH NOVELIST, POET,
PLAYWRIGHT, ESSAYIST, LITERARY CRITIC AND PAINTER

Alongside concerns over badger culls and the impact of neonicotinoid insecticides on bees and other pollinators, there is another farming issue that has hit the headlines in recent years: the price of milk. Not its escalating price, quite the opposite: concern that the lower and lower prices seemingly 'demanded' by consumers and the dominant supermarkets are putting many smaller dairy farmers out of business. In 1996 there were 34,570 dairy farms in the UK; by 2015 that had reduced to fewer than 13,355.[i] A survey in the autumn of 2016 by AHDB Dairy (the levy-funded, not-for-profit organisation working on behalf of Britain's dairy farmers) found that nearly 11% of our dairy farmers intend to exit dairying within two years.[ii]

Farmers' margins on milk have slumped over the years, partly because milk is now traded as a global commodity so prices are determined in a world market, and partly because consumers want cheap food provided mainly by the large – and obliging – supermarkets. According to AHDB Dairy, by 2016 world market prices were showing signs of a little recovery. Since farmers can't rapidly convert in or out of dairy farming – it takes a few years to assemble a milking herd and install equipment – this might suggest more than a temporary blip and herald the first signals of some good news for dairying.

While many dairy farmers have given up on milk as a result of low profit margins, others have become much bigger, milking larger numbers of dairy cows to get the economies of scale, and converting almost all of their pasture to intensively-managed, regularly-fertilised, highly-productive grass leys. The average size of a UK dairy herd in 1996 was 75 cows; by 2015 that had risen to 142.[i] Farms with more than 200 milking cattle are today commonplace; many are much larger still, and dairy units housing thousands of cattle indoors all year are being contemplated. Production is becoming increasingly concentrated in the west of England, mainly Devon, Somerset, Cheshire, and Cumbria. There is also a significant movement in volume from England to Wales, Scotland and Northern Ireland, indicating increasing production in these countries.

Although most dairy herds are grazed on pasture all summer, they spend a long winter in huge cattle sheds where they are fed silage cut on the farm the previous summer, and feeds bought in from off the farm. Some farmers feed their dairy cows indoors all year, reducing the effort of gathering them up for milking twice a day. With milk prices at an all time low, getting high yields from each cow is often vital for economic survival. In 2005 the average UK dairy cow produced 7,000 litres of milk a year; by 2015 that had become nearly 8,000 litres.[iii]

Gathering in cattle for milking from verdant green pastures epitomises what most people consider a traditional farming scene in the British lowland countryside. But the continuous process of increasing milk yields means that most of the pastures they graze have changed enormously in the last few decades. Permanent pastures in which common flowers like buttercups, Red Campion, forget-me-nots, and stitchworts added specks of colour to the grasses have largely been replaced by much more productive (for grazing dairy cattle) temporary pastures. These leys are ploughed up and re-sown every few years with a mix of fast-growing ryegrasses and clovers. The wildlife value of fields converted from permanent pasture to grassy leys plummets; its plant diversity is very limited and so too are its invertebrates and everything else up the food chain.

But not all dairy farms are large and not all dairy farmers are managing their land as intensively as they might. Smaller units can still make a profit if they produce most of their winter cattle feed on the farm rather than buying it. Great Wollaston Farm near Shrewsbury is a good example. A long-established LEAF (Linking Environment and Farming) farmer, Robert Kynaston milks 80 Friesian and Holstein crosses and produces around 650,000 litres of milk a year. He sells to Arla, a farmer-owned international milk company. He has 45 hectares of pasture (about a third of which is permanent pasture) and he grows peas, wheat, and barley on another 40 hectares. His cows are kept indoors from November until maybe March and he is virtually self-sufficient in winter feed for them.

'If I wasn't almost self-sufficient in feed, it would be difficult to make a profit,' says Mr Kynaston. 'I grow spring-sown peas and barley plus winter wheat and barley on about 40 hectares of arable land in a rotation with temporary grass and clover leys. I have to buy in some compound feed for them in winter but that's all. I incorporate clover in the leys and don't use much fertiliser and I'm an advocate of minimum tillage so we do no deep ploughing here. We're getting about 19 pence a litre for the milk at the moment (2016) but we're profitable.'

CRÈME DE LA CRÈME

The UK's place in the international milk stakes:

- The UK is the third-largest milk producer in the EU after Germany and France, and the tenth-largest producer in the world

- Milk accounted for 17.8% of total agricultural output in the UK in 2014, and was worth £4.6 billion in market prices

- The total number of UK dairy cows has fallen from 2.6 million in 1996 to 1.9 million in 2016 (though it has been increasing slowly over the last five years), a 27% reduction, while herd size has increased

- The UK produced 14.6 billion litres of milk in 2014, little more than in 1975, but the milk yield per cow has doubled in that time

- In 2014, the UK had a negative trade balance in butter and cheese, but a positive trade balance in milk and cream

- UK farm-gate milk prices for November 2016 averaged 25.6 pence per litre

- Between 1995 and 2015, doorstep delivery has declined from 45% to 3% of the retail milk market

(Source: UK Dairy Industry Statistics, Alex Bate, 2016. House of Commons Briefing Paper no. 2721.)

However, there is an alternative to grass and clover leys for many farms feeding a herd of dairy cows. Chapter 6 referred to the advantages of substituting them with herbal leys. Sown with a mixture of grasses such as Cocksfoot, Timothy, and Meadow Fescue, plus legumes including Red Clover, Alsike Clover, Birdsfoot Trefoil, and deep-rooting forage herbs such as chicory, Sainfoin, Great Burnet, and Sheep's Parsley, their establishment cost is higher but they require no expensive fertiliser. There is also evidence that they naturally reduce livestock gut parasites.

'There's a significant interest in establishing herbal leys as an alternative,' comments Ray Keatinge, Head of Animal Science at the AHDB. 'For grazing livestock, many of the plants in the mix are deep rooted so they draw up minerals from deeper down in the soil so stock seem to do better overall than they do on ryegrass-clover mixes. Herbal leys are more drought resistant and that's likely to be increasingly important particularly in the south of England if the climate warms further and where soils are thin. Not requiring fertiliser is a big cost advantage and the claims that some of the plant species protect against gut parasites is a bonus too. Herbal leys don't yield as much vegetation as a ryegrass-clover mix

New herbal? Better for dairy cattle, better for wildlife; could herbal leys replace wildlife-poor ryegrass pastures? (courtesy of Cotswold Seeds).

and the seed cost for a herbal ley is more but yield isn't everything. Soil and water management – and their wildlife value – are important considerations too,' he adds.

There is growing evidence that herbs, such as forage chicory and plantain, can be high-yielding and beneficial sources of very palatable and nutritious feed for grazing livestock.[iv] Using species like these can reduce livestock farmers' reliance on concentrate feeds. Although chicory has been used in agriculture for some time, its use as a forage crop for livestock is relatively new. Much like chicory, plantain has only recently established itself as a viable forage crop for livestock production, and very few fields of it have been grown in England to date. Most evidence of the potential of these plants lies in the southern hemisphere, where modern varieties of both forage chicory and plantain have been developed. Both have large tap roots, making them potentially very valuable in a drought. Moreover, increasing the diversity of plant species in grazed pastures provides a better habitat for a wider range of invertebrates than pastures restricted to a couple of fast-growing grasses and clover.

Professor Chris Reynolds, Director of the Centre for Dairy Research and his colleagues in the University of Reading's School of Agriculture, Policy and Development, was awarded a substantial grant in 2015 – alongside further funding for collaborators at Rothamsted Research – to explore how diverse forage mixtures could optimise the production of milk and meat from ruminant animals such as cattle and sheep. The idea was to see if such developments improve the efficiency of animal feed production, reduce the environmental impact of UK cattle and sheep farming and its dependency on feed imports, and boost wildlife on farms.

In several other countries, grazing dairy cattle on naturally flower-rich pasture is the norm. Ask a farmer in Austria why he doesn't plough up his valley-bottom, flower-burgeoning pastures and sow them with a ryegrass and clover mix and you will get a rather shocked response. The farmer might be part-time and his farm might be no more than 20 hectares at most, but he wants his dairy cows to produce quality milk to raise his calves; having pasture rich in a wide variety of plants is just the ticket. He will cut silage off a couple of his pastures to feed the cattle indoors over the long winter but he won't take more than two cuts a season because he knows that too regular cutting will transform his flowering pasture into a grasses-only field. He certainly doesn't want that. Theirs is the natural version of a herbal ley; it's teeming with invertebrates all summer because of its huge variety of flowering plants.

In the UK, farmers have used CAP money to convert most of our traditionally-managed hay meadows and pastures into more productive leys, either by ploughing them up and reseeding them with a ryegrass/clover mix, or by fertilising them over the years which gives grasses a competitive advantage while other plants slowly die out. Herbal leys don't turn the clock back and re-introduce the native plants that have largely died out but they do introduce a much welcomed diversity of plants into the grazed sward. Laurence and Tom Harris at Ffosyfìcer dairy farm in Pembrokeshire (see Chapter 10) intend to try some herbal leys in the future. For an organic farm like theirs with 350 milking cows producing around two million litres of milk a year, the attraction of this nutritious feed is twofold: an increase in their milk yields, and potentially lower vet and medicine bills through its ability to reduce gut parasites in cattle.

Large dairy farms produce a huge amount of manure which, if not properly stored and used, can cause enormous pollution problems if it gets into ditches, streams, lakes, ponds, and rivers, killing off aquatic invertebrates and thereby reducing the survival chances for most fish, small mammals, and stream-feeding birds left without their prey. Nitrogen applications (mainly using synthetic fertilisers) to grassland have declined markedly over the last two decades but remain about the same on arable crops. Phosphate applications have declined both on grassland and on arable land over the same period, though they have levelled out in the last few years. Farmers are being generally more frugal with expensive fertiliser, using only the quantity they really need. Nevertheless, farming accounts for 60% of the nitrogen and well over a quarter of the phosphorous getting into our rivers,[v] much of it from manure runoff.

The dairy industry contends that 77% of dairy holdings in 2015 were implementing nutrient management plans and that most are updating their plans every year, thus matching nutrient inputs (from fertilisers and manures) to crop demand so that losses in runoff are minimised.[vi] That's borne out by a decline in pollution incidents. Between 2013 and 2014, reported serious pollution incidents caused by dairy farms decreased by 18%, down from 44 incidents to 36 UK-wide. Longer term (2008 to 2014), while the number of dairy farm pollution occurrences has risen in some years, overall there has been a 23% drop in incidents caused by the dairy sector.

But there is no room for complacency, and the farming industry generally needs to be more proactive in curbing fertiliser and manure runoff from farmland because many water bodies remain polluted. There was a decrease, albeit a slight one, in the overall number of water bodies awarded high or good surface water status between 2011 and 2016 from 37% to 35%, with pollution from agriculture remaining a significant cause of problems.[vii]

Based on its phosphate content, the manure output of 100 cows roughly equates to that produced by 1,400 people.[v] About 96 million tons of farm manures – solid manures and slurries – are applied to agricultural land each year in the UK. Much of it is from dairy units but the intensive rearing of pigs and chickens, many of which are kept at least part of the time in massive indoor units, also produces huge quantities of manure that have to be removed and handled. When properly managed, dairy and other livestock waste makes an excellent fertiliser, promoting crop growth, increasing soil organic matter, and improving overall soil fertility and tilth characteristics.

FREE FERTILISER

The original – and free – pasture and crop fertiliser, manure is still the mainstay of organic farming which shuns synthetic fertilisers. Farmyard manure isn't solely the faeces of livestock; it also contains plant material, often straw, which has been used as bedding for animals and has absorbed their faeces and urine. Liquid form – known as slurry – is produced by more intensive livestock rearing units in which concrete or wooden slats are used instead of straw bedding.

Manure from different animals has different qualities and requires different application rates when used as fertiliser on pasture and crops. Sheep manure is high in nitrogen and potash; pig manure is relatively low in both. Chicken litter is very concentrated in nitrogen and phosphate and is prized for both properties.

Animal manure spread on land encourages soil microbial activity because it provides food for a huge range of soil animals and fungi. In turn this promotes the soil's trace mineral supply, improving plant nutrition while its nitrogen and other nutrients stimulate plant growth.

Manure running off land into watercourses isn't the only way in which it can cause pollution and damage wildlife. Ammonia – formed by the breakdown of urea in mammal manures and the breakdown of uric acid in chicken manure – escapes as a gas into the atmosphere, especially when manure or manure slurry is spread on farmland. In 2014, farming accounted for almost 83% of all ammonia emissions in the UK, and although they fell overall by a quarter between 1990 and 2014 due to reductions in cattle numbers and more efficient fertiliser use, they started rising again between 2013 and 2014.[v]

The problem is that once it's up in the atmosphere, ammonia can travel on the wind and be deposited in rainfall distant from where the manure was originally spread. When it comes back down to ground it can change the plant composition of habitats that depend on low levels of nitrogen (ammonia is a rich source of nitrogen), stimulating plants like grasses that respond well to its fertiliser effect rather than plants such as heathers and heaths that don't. Deposited ammonia can also cause soil acidity to increase, again changing the vegetation composition of habitats, for instance by slowing the growth or killing off some lichens and mosses.

Just 4% of the total ammonia emissions from a farm are deposited within 200 metres of the farm; the remaining 96% travels further, often much further, depending on weather conditions. In a study of the impact of an intensive poultry-rearing unit in North Wales housing up to 180,000 birds, ammonia emissions were detected up to 2.8 kilometres upwind, contributing 30% of the total nitrogen deposited in parts of a large coastal sand dune habitat of national importance where its long-term impact is unknown.[viii] Locating several such intensive rearing units in close proximity increases the possibility of damage to nearby habitats from ammonia escaping to the atmosphere.

One of the first steps to lower levels of ammonia is to ensure that livestock are not fed more protein than is necessary, thereby reducing the amount of nitrogen excreted as urea in urine and faeces – or uric acid in poultry – which would be mostly converted to ammonia. Increasing the efficiency of nitrogen use is more difficult when farming cattle and sheep, although supplementary feeding using high energy, lower protein feed – such as maize silage – might be beneficial. It's also important to keep livestock buildings clean; ammonia losses are higher if the walls and floors are constantly covered with urine or layers of faeces. For poultry, ammonia loss can be reduced by ensuring that their droppings are able to dry rapidly.

Reducing ammonia loss from slurry stores is also essential. In 2016 Defra announced the Farming Ammonia Reduction Scheme in England, in which farmers can apply for a share of a £15.9 million pot to install slurry store covers as part of an EU initiative. Slurry stores need to be covered to stop rain getting in, diluting the slurry but increasing its volume, therefore running the risk of overflow. The funding is available only for beef and dairy farmers, but it's reasonable to ask why it is that any significant slurry store has not been subject to compulsory covering before now and why farmers don't always consider it good practice to do so at their own expense.

Accurately applying manures and slurries to land when it is spread is also important in reducing ammonia emissions, the most effective way being to inject it directly into the soil as Tom Harris does at Ffosifycer. Spreading slurry on a cool, still day rather than a warm or windy one keeps emissions down too. Applying it just before or during rainfall can also be beneficial – if the soil is free-draining and there's no risk of causing water pollution in nearby watercourses – because it can wash quickly into the soil.

Instead of using farm animal manures as a fertiliser, a few big farms, mainly pig and poultry units, have opted for a very different handling solution: anaerobic digestion. Popularly known as AD, it's the controlled breakdown by bacteria of organic matter without air to produce a combustible gas and a nutrient-rich organic by-product that is ideal for land spreading. The gas – a mix of methane, carbon dioxide, and other gases in small amounts – can then be used on the farm to generate electricity or for running a heating system. The organic by-product or digestate is a low odour, slow release material that is easier to handle and spread than manure, and adds a lot of nutrients including nitrogen and phosphate to the soil. The financial viability of this technology depends on its scale; a dairy farm would need at least a few hundred cows to justify an AD system but for many larger pig and poultry units they make good economic sense. The main environmental gain is that they reduce greenhouse gas emissions (mostly methane) to the atmosphere when raw manure is field spread; they also reduce the volume of manure so that the chances of large amounts washing off fields and polluting watercourses are lessened too.

So what regulations are in place to make sure farmers take the utmost care if they are spreading manure or manure slurry as fertiliser on their fields? In order to receive their Basic Payment Scheme (BPS) subsidies, farmers in England are required to produce and keep a map of the farm if they spread manure. It has to show all land within ten metres of any surface waters on the farm plus all springs, wells, and boreholes and land within 50 metres of these.[ix] Farmers must not apply manure within these limits, although with certain forms of application such as direct injection into the soil the manure-free buffer strip can be decreased from ten to six metres. A farmer can be subject to an unannounced spot check by the Rural Payments Agency at any time and small reductions in BPS payments can result from non-compliance. Proposals to toughen up financial penalties are being suggested, with some politicians proposing that subsidy payments should be withdrawn from a farmer found to be polluting watercourses. 'This is corporate crime. If you damage the water with slurry, or the air by releasing ammonia, you should lose your subsidies,' commented Mary Creagh, Chair of the House of Commons Environmental Audit Committee in August 2017.[x]

More stringent rules to avoid the pollution of watercourses apply within Nitrate Vulnerable Zones (NVZ's), areas of land where it's more likely that water bodies and drinking water supplies will get polluted by nitrates from farming. The rules are complex (many farmers claim they are far too complex) and include: controlling the dates and conditions under which nitrogen fertiliser and organic materials are spread; having sufficient on-farm facilities for storage of manures and slurries; limiting nitrogen fertiliser applications to the crop requirement only; restricting quantities of organic material (from any livestock including pigs and poultry) applied per hectare per year; regulating the total quantity of organic material applied to the whole farm; managing the areas where nitrogen fertilisers (both organic and inorganic) can be applied; organising application methods; and preparing plans and keeping adequate farm records.[xi,xii] They are enforced by the Environment Agency in England and its equivalents in the other UK countries. The UK Government is also considering insisting on crop or vegetation cover on all soils at all times of year to reduce slurry and fertiliser runoff during periods of heavy rain.

THOSE VULNERABLE ZONES

Nitrate Vulnerable Zones (NVZs) were introduced in 2008 by the EC Nitrates Directive with the aim of reducing water pollution caused by nitrates from agricultural sources, including manures. As more of the farmed land in England is lowland, and because much of that receives far less rainfall than the uplands (so any pollution is less likely to be naturally diluted), around 60% of land in England is included in NVZs.

The proportion in the other UK countries is much lower; in Wales it's 2.4% of the land area, although discussions are underway about increasing the amount of land included in NVZs, especially in the lowlands of Pembrokeshire where there is a significant amount of dairy farming. The Welsh Government launched a consultation in September 2016 suggesting either a targeted approach in particularly vulnerable areas of Wales that drain into currently polluted waters, increasing the NVZ area from 2.4% to 8%, or making all of Wales an NVZ as has been done for the whole of Northern Ireland![xiii]

The Welsh Government argues that:

> ...a whole Wales NVZ would provide an opportunity to develop an integrated approach to a number of different, though related issues. As well as providing a baseline to tackle agricultural pollution it could also help us to meet other key objectives on air quality and reducing greenhouse gas emissions. Using a single legal instrument to set a baseline level of farm nutrient management would give us the opportunity to coordinate action aimed at achieving our objectives and make the baseline rules as simple as possible.[xiii]

Blooming nuisance. Soil laden with fertiliser washing into waterways
can cause algal blooms and kill aquatic animals (courtesy of François Lamiot).

Action is clearly needed: between 2010 and 2015 in Wales, an average of 61% of recorded agricultural pollution incidents per year were from dairy farms and 19% from beef farms. What's more, slurry stores on the majority of farms are too small or too old to be capable of storing enough manure slurry through the winter months when spreading will not be possible within an NVZ because this is the very time (especially in the west of the UK) when rainfall is likely to wash it into watercourses and cause pollution. As a consequence, many farmers and contractors are pressing the Welsh Government for flexibility in slurry (and fertiliser) spreading dates rather than a fixed closed period which would rule out any spreading in dry periods of autumn weather.

Farmers have to keep a number of records for land in an NVZ and that's probably why it's one of the most common areas of cross-compliance failure that results in fines. NVZ farmers must have a risk map of each field, its soil types, its slope, the locations of manure heaps, details of crops being grown, and quantities of manure or fertiliser spread on them, plus overall farm data on cattle numbers and the quantity of nitrogen in the manure they produce. Its critics argue that there's too much emphasis on checking that farmers have the correct paperwork, rather than checking on what practical results they have achieved. A dishonest farmer can have all his paperwork in order but might still be polluting streams.

Leaving the EU will mean that the UK is not obliged to implement any EU Directives but, because of the seriousness with which environmental pollution is viewed in the UK, it's unlikely that measures such as those in the Nitrates Directive will be abandoned. Once we have left the EU, it might also be the case that remaining EU countries will not agree to trade in farm produce with the UK unless our farmers continue to adhere to EU standards. Relaxation doesn't seem to be an option. Until appropriate new legislation is considered by the UK and devolved governments, it seems likely that the requirements of the Directive will be transposed directly into UK legislation and may well stay that way. However, the system does need to be more focused on practical reality and not paper records.

As well as the impact of ammonia emissions, runoff from manures (and excess synthetic fertiliser) spread on pasture and arable land can have a catastrophic impact on habitats distant from the farmland on which they were applied. In ditches, streams, rivers, ponds, and lakes, nutrient pollution – mainly nitrogen and phosphate – from fertilisers and manures causes eutrophication and sometimes considerable ecological damage.

BLOOMING TROUBLE

Eutrophication is a process in which excess nitrate or phosphate (or both) in the water encourages the growth of algae that forms a bloom over the water's surface. This bloom prevents sunlight reaching other water plants which then die. Bacteria break down the dead plants and use up the oxygen in the water so the lake, ditch, or stream may be left completely lifeless. Aquatic invertebrates die off rapidly and most fish can't survive; aquatic birds such as Yellow Wagtails and Dippers that rely on invertebrates can't find any prey while Kingfishers are deprived of their fishery. Water plants that are acclimatised to naturally low nutrient conditions are likely to be killed off too, and the effect of this pollution can frequently be felt for many kilometres downstream and for prolonged periods, often semi-permanently, in landlocked water bodies such as ponds and lakes.

Undiluted animal manure or slurry is 100 times more concentrated than domestic sewage while silage liquor (from fermented wet grass) is one of the most polluting organic liquids. Eutrophication can often be worse in winter after autumn ploughing has released a surge of nitrates; winter rainfall is heavier, thus increasing runoff and leaching, and there is lower plant uptake of these nutrients because plant growth has slowed in the cool conditions.

It's little wonder then that farmers are encouraged to develop manure and fertiliser management plans for their farm. There is plenty of advice available to help them do so, covering a range of measures: mapping areas close to water bodies where fertilisers and manures should not be spread; categorising land into high, moderate, and low risk zones; taking account of sloping land where runoff can be greatest; calculating crop nutrient requirements so that excess fertiliser or manure is not spread on land; and limiting losses to the atmosphere by adopting better management such as direct injection into soil rather than spreading or spraying.

Injecting slurry also ensures that the soil gets all the nutrients from the manure whilst limiting the amount of nitrogen released as ammonia into the atmosphere. It's also why farmers have to keep any soil bare of plant growth for as short a time as possible, leaving winter cereal stubbles in place all winter or sowing quick-growing cover crops to blanket the soil before a main crop can be sown. These measures have a significant wildlife spin-off value; there are more opportunities for native plants to grow and set seed – and habitat for invertebrates – while crop stubbles are excellent feeding places for small birds (see Chapter 7).

The huge fall in milk prices over the last decade has not only driven many smaller dairy farms out of business, it has also had a hugely detrimental impact on wildlife too. The perfectly natural response of many farmers, working long and hard for little profit, has been to intensify their operations. That means more cattle on more pasture land that is more grass productive. More grass-productive pastures are attained by ploughing and re-seeding with grass/clover leys or boosting soil fertility, some of which can be cut four times a season to provide silage for winter cattle feed. Boosting grass growth elbows out virtually all other meadow and pasture plants; common flowers such as hawkbits, Yarrow, Yellow Rattle, and Sweet Vernal-grass have become very hard to find except where livestock are kept away from well-maintained hedges and a few other less-accessible spots. Little wonder that such pastures are poor for invertebrates and the only birds that sometimes feed on them might be a few crows, some passing gulls, and maybe a few Starlings.

But there are alternatives. Grazing dairy cattle on herbal leys instead of ryegrass/clover mixtures is one. More expensive to establish but requiring no fertiliser, providing a healthier diet for cattle and coping better in drought, they can be cheaper overall and attract a wide variety of invertebrates. Organic dairy farming is another alternative as long as product markets can be established; if so, the land management without synthetic pesticides and fertilisers encourages more plant diversity in grazed pastures, in turn creating a better habitat for a range of invertebrates, small mammals, and birds.

The future for the UK dairy industry outside the EU will remain driven by world milk prices. It also seems likely that a post-CAP future of reduced or non-existent BPS subsidies will stimulate further enlargement of dairy farm units to achieve greater economies of scale. It will therefore become yet more important that environmental protection measures including NVZs remain in place because of the continuing, and perhaps increased, risk of serious pollution incidents affecting streams, rivers, or other water bodies. Dairy farmers need to be encouraged via agri-environment schemes to invest in establishing herbal leys to replace ryegrass/clover pasture wherever possible and to adopt the safest forms of on-farm manure and slurry handling. Some organic dairy farmers have established good markets for their products and governments across the UK should give more encouragement to dairy farmers to convert to organic production because of the benefits it provides for wildlife.

Endnotes

i dairy.ahdb.org.uk/.../farming-data/average-herd-size.
ii 'How might farms' future plans affect GB's milk profile?' AHDB Dairy, 19 October 2016.
iii 'Dairy Statistics: An Insider's Guide, 2016.' Agriculture and Horticulture Development Board Dairy.
iv 'Using Chicory and Plantain in Beef and Sheep Systems,' Dylan Laws and Dr Liz Genever, 2013. Agriculture and Horticulture Development Board.
v ' Agriculture in the UK 2015,' Defra and devolved Governments, 2016.
vi 'The Dairy Roadmap 2015,' AHDB Dairy, Dairy UK, the National Farmers Union, and others.
vii B7. Surface water status. UK Biodiversity Indicators 2017. JNCC, 2017.
viii 'Upwind impacts of ammonia from an intensive poultry unit,' L. Jones etal, 2013. *Environmental Pollution* 180: 221- 228.
ix 'Guide to Cross Compliance in England: 2016,' Rural Payments Agency, 2016.
x 'Outrage over unfair pollution allegations,' *Farmers Weekly*, 25 August 2017, pages 6–7.
xi 'Guidance on complying with the rules for Nitrate Vulnerable Zones in England for 2013 to 2016' Defra, 2013.
xii 'Nitrate Vulnerable Zones in Wales Guidance for Farmers,' Welsh Government, 2014.
xiii Review of the Designated Areas and Action Programme to Tackle Nitrate Pollution in Wales. Welsh Government Consultation document WG27622. 2016.

CHAPTER 13

Evolution, revolution

*'Go back?' he thought. 'No good at all! Go sideways?
Impossible! Go forward? Only thing to do! On we go!'*

J.R.R. Tolkien (1892–1973), *The Hobbit*

Farming has been substantially altering our landscapes and their wildlife for millennia. Chapter 2 summarised the enormous changes that have taken place since the first farmers arrived here sometime before 6,000 BC and modified the means and intensity of farming, the land it occupies, and the wildlife associated with it. Over the last few decades, especially in the lowlands of Britain, farming has transformed as a result of better crop and livestock breeding, the advent of a range of pesticides to help control plant diseases and crop weeds, and the use of synthetic fertilisers. It has become a very much more intensive, almost industrialised operation, with crop and milk yields and livestock productivity at levels considered impossible even half a century ago.

More change is yet to come. One of the debates about its future direction – perhaps the most controversial – revolves around GMOs: genetically modified organisms. Should UK farmers grow genetically modified (GM) crops? Are they a way of reducing the use of some pesticides and increasing crop yields or will public opinion remain opposed when scientific opinion is supportive?

There are other issues too. Will climate warming necessitate abandoning some of the crops we grow now and introducing others, especially in southern Britain? Are farmers responding to the likelihood of more frequent droughts? Is public concern about issues such as animal welfare, the decline of pollinating insects, and wildlife declines in general going to stimulate changes to farm supports? Are we going to get serious about using our hills and moors in the north and west of

Britain much more effectively as water sponges in order to help alleviate the cost and misery that increases in rainfall and therefore flooding can cause?

Overshadowing all of this, at least for the next few years, is Brexit. Leaving the EU means leaving the Common Agriculture Policy (CAP), which has formulated farming policy and governed its direction for as long as many of today's farmers can remember. When farming policy is 'repatriated' to the UK, how it's supported and what strategies might be established to guide its direction will be decided by our four different governments. Complicated, maybe. But a vote at some point in the future for Scottish independence could even return one nation of the UK back to the CAP fold as an individual EU member. More complicated still.

Of the potential issues the farming industry needs to grapple with, GM crops could be one way of reducing pesticide use on farms, benefitting many plants, invertebrates, birds, and small mammals. However, public opinion in the UK remains opposed to their introduction here, in spite of imported GM produce being consumed in several foods. Only one GM crop – a type of maize used for animal feed – is grown commercially in the EU. There have been no new GM crops approved by the EU since 1998, apart from a type of GM potato which was later withdrawn, as countries opposed to GM had, until 2015, been able to block Europe-wide approval.

The area planted with GM crops worldwide increased from 1.7 million hectares in 1996 to 181.5 million hectares in 2014, 13% of the world's arable land, with an increasing proportion grown by developing countries. The largest growers in order are: USA, Brazil, Argentina, India, Canada, and China.[i] By 2015, 92% of corn, 94% of soybeans, and 94% of cotton produced in the US were genetically modified strains.[ii] Nine countries have opposed their cultivation in the EU. While in some EU countries GM foods are seen as a potential threat to the reputation of its agricultural produce, in others, governments argue they are essential to feed a growing world population.

No GM crops are currently being grown commercially in the UK, although imported GM commodities, especially soya, are being used in animal feeds and to a lesser extent in some food products. For the UK, Defra has set out the procedure for anyone who wants to release a GM organism or market a GM product.[iii] It makes clear that applications for approval to market a product (including crop seeds for cultivation, foods, or feeds) are assessed and decided currently at EU level and involve the European Food Safety Authority, while applications to release a GM organism for research and development purposes are considered at national level. The assessment process considers potential safety factors such as toxicity, allergens, and the fate of any possible transfer of novel genes to other organisms such as wild plants. Applicants have to provide a dossier of relevant information to cover these points, and this is scrutinised by the UK's independent Advisory Committee on Releases to the Environment (ACRE). In 2015 these arrangements were modified to give final responsibility for local implementation back to member states who can now decide whether to opt out from cultivation of a GM crop that might be authorised at EU level.

IT'S IN THE GENES

GM technology enables plant breeders to bring together in one plant useful genes from a wide range of living sources, not just from within the crop species or from other plants, which traditional plant breeding has long done. It speeds up the process of generating superior plant varieties, and it expands the possibilities beyond the limits imposed by conventional plant breeding

The first genetically modified crop plant was produced in 1982, an antibiotic-resistant tobacco, and the first field trials occurred in France and the USA in 1986 when tobacco plants were engineered for herbicide resistance. The first pest-resistant tobacco plants were genetically engineered in 1987 by incorporating genes that produced insecticidal proteins from a naturally-occurring soil-living bacterium, *Bacillus thuringiensis* (Bt).

The People's Republic of China was the first country to allow commercial growing, introducing a virus-resistant tobacco in 1992. The first genetically modified crop approved for sale in the US was the *FlavrSavr* tomato in 1994. It had a longer shelf life because it took longer to soften after ripening. In the same year the EU approved tobacco engineered to be resistant to the herbicide bromoxynil, making it the first commercially genetically engineered crop marketed in Europe.

In 1995, Bt potato was approved by the US Environmental Protection Agency, making it the country's first pesticide-producing crop. The same year, canola (a variety of rapeseed) with a modified oil composition, Bt maize, bromoxynil-resistant cotton, Bt cotton, glyphosate-resistant soybeans, and other crops were all approved. By 1996, a total of 35 approvals had been granted to commercially grow eight genetically modified food crops and one flower crop (carnation), with a number of different traits. In 2000, vitamin A-enriched golden rice was developed, though it's not yet in commercial production. In 2013 the leaders of the three research teams that first applied genetic engineering to crops, Robert Fraley, Marc Van Montagu, and Mary-Dell Chilton, were awarded the World Food Prize for improving the 'quality, quantity or availability' of food in the world.

Trials of GM crops have been licensed in the UK, often provoking strong objections and active interference from advocates of GM-free farming. In November 2016, researchers applied to Defra to get consent to grow a GM wheat in the specially-fenced GM-dedicated growing area at Rothamsted in Hertfordshire, the longest-running agricultural research station in the world. The plants have a gene added from a grass called Stiff Brome which researchers believe will produce much higher-yielding wheat.

Any new technology brings risks. With GM crops there are concerns about the chances of modified genes escaping from cultivated crops into wild relatives, the likelihood of pests evolving resistance to the toxins produced by GM crops to kill them, and the possibility that these toxins could affect non-target organisms.

According to the Royal Society there is no evidence that producing a new crop variety using GM is more likely to have unforeseen effects than creating one through conventional cross-breeding.[iv] They go on to say:

> Concerns have been expressed that simply inserting new DNA into a plant genome by GM, might have unpredictable consequences. However, as our knowledge of genomes has increased it has become clear that similar insertion events occur frequently in all plants. For example, some bacteria and viruses insert new genes into the genomes of plants that they infect. We also know, from studying the genomes of different members of the same species, that gain and loss of genes within species is very common too. GM crops are more extensively tested than non-GM varieties before release.

But GM's critics are not as sanguine. Friends of the Earth, for instance, remains to be convinced that GM crops incorporating a toxin which kills the crop's pests will not affect other insects such as butterflies and moths that might also consume the toxic pollen.[v] It's why they are concerned about growing GM maize which incorporates a gene that produces a poison to kill European Corn Borer Moths, a particular pest of the crop. There is evidence that other butterflies and moths which consume maize pollen might be killed too. In addition, there is some concern that widespread growing of the crop could lead to its insect pests developing resistance to the contaminant.[vi] The advantage of GM maize is that less insecticide spraying is necessary, thereby benefitting a wide range of insects. Eventually, decisions might have to be taken on a balance of probability, weighing up the advantage of less insecticide against the possible impact of the GM crop on non-target insects.

Another standoff concerns GM potatoes and a fungal disease – potato blight – prevalent in damp climates like those in the west of Britain. It was the disease that caused the horrendous famine in Ireland in the 1840s. A no-blight GM potato has been developed (incorporating genes from particularly disease-resistant varieties), crops of which would not need to be sprayed several times a season. Some potato crops are currently sprayed as many as 20 times a season with a fungicide. The crop is also resistant to nematode worms and to bruising in transport. Using the GM spud would stop the sprays killing off a multitude of beneficial fungi which perform critical roles in soil ecological processes, including the essential mycorhizzal fungi that garner soil nutrients for crop roots. Critics argue that there are many blight-resistant varieties of potato already available to growers, all of them developed by normal plant breeding rather than resorting to GM. The result is that the GM potato – inevitably dubbed 'the super spud' – has not been given the go-ahead for planting in the EU.

Top 10 Countries with Commercial GM Crops (2015)

Country	Cultivation area (Hectares)	Crops
USA	70.9 million	maize, soybean, cotton, canola, sugarbeet, alfalfa, papaya, squash, potato
Brazil	44.2 million	soybean, maize, cotton
Argentina	24.5 million	soybean, maize, cotton
India	11.6 million	cotton
Canada	11.0 million	canola, maize, soybean, sugarbeet
China	3.7 million	cotton, papaya, poplar
Paraguay	3.6 million	soybean, maize, cotton
Pakistan	2.9 million	cotton
South Africa	2.3 million	maize, soybean, cotton
Uruguay	1.4 million	soybean, maize

(Adapted from data compiled by GeneWatch UK.)

Another potential environmental problem referred to by GM critics is that GM modified crops could breed with closely-related native plants, creating hybrids. Such hybrids do occur naturally of course. Spanish Bluebells for example, introduced via gardens into the UK, have been hybridising with the English Bluebell, resulting in concerns from conservation organisations that the long-term survival of the pure English Bluebell is in doubt. To date, there's been no recorded instance of a GM plant hybridising with a native plant.[vii] If it did occur, due to the fact that most agricultural crops are not native to the place they grow in, the crop might have few related wild species near at hand and a hybrid might not be well adjusted to survive in the wild anyway.

The criticisms of GM crops are not accepted by leading plant scientists, academics, and many others, who argue that the gains far outweigh any concerns. In a recent series of essays the authors criticise what they refer to as a 'prolonged and shallow debate' about GM crops in Europe during the past two decades, imposing great costs on farms and the environment.[viii] Compared with farming in many other parts of the world where GM technology is proving advantageous, they argue that European farmers are being left behind.

Spot the difference. Genetically modified potatoes, known as Amflora potatoes, modified to produce pure amylopectin starch (courtesy of BASF Plant Science).

It isn't an argument bought by the Soil Association. 'It's familiar, old propaganda,' commented Peter Melchett, the Association's Policy Director, when the document was published. 'There is not a single voice from food retailers or caterers calling for GM food, even after the 20 glorious years that this report claims for this outdated and failing technology.'

Might there be other advantages for technologies such as GM crops? Some academics claim that if farmers embraced up-to-date technologies such as GM to greatly increase crop yields, more land could be set aside for tree planting which would act as a greenhouse gas store. It's a 'land sparing' argument that's been around for a long time but which might be more applicable to countries with large land masses where vast areas of intensive agriculture (almost without any wildlife habitat) contrast with other land areas set aside strictly for wildlife. In the UK, with our more intimate patchwork of farmed land and bits of wildlife habitat – and often more fragmented land ownership – it's very difficult to implement. Also, the British philosophy has traditionally been a more 'we can muddle along together' approach in which compromise is often the order of the day!

Whilst EU member states are able to restrict or even prohibit the cultivation of GMs in their territory, the EU does not permit a general ban on GM cultivation; the restrictions or prohibitions apply only to a specific GM crop. In the UK, decisions on growing GM crops fall to the respective devolved governments with Defra representing the UK Government and they are expected to reach an agreed UK-wide position. If they have differing views, Defra effectively has the 'casting vote'. Post Brexit, the differing approaches to farm policy already apparent within the UK's constituent nations are likely to be increased; we wait to see whether each devolved administration argues that it should have the authority to decide whether to authorise a GM crop within its own territory.

Environmentally advantageous or not, it's probable that continuing public concern, much of it ill-informed, that GM crops are not safe to grow and eat will determine whether any are grown on UK farms in the future. It seems unlikely that GM crops will become a feature of the UK's agricultural landscape for some time. Eventually though, as more and more countries make use of them, and if no significant environmental, wildlife, or nutritional ill-effects are discovered, our farmers will be growing some of them too.

––––––––––––––

Although coal and gas power stations, industry and transport are the major sources of greenhouse gases linked to climate warming, it's easy to forget that agriculture contributes to it substantially too. Agriculture produces 9% of the UK's greenhouse gas emissions.[ix] Over half of it is nitrous oxide produced by synthetic and organic fertilisers applied to farmland. Agriculture is the largest human source of the gas, accounting for approximately three-quarters of our nitrous oxide emissions. Another third or more of agricultural emissions is methane created by the digestion processes of livestock animals – what Cockney rhyming slang calls 'raspberry tarts' – as well as the production and use of manure and slurry. Cows produce the most methane per animal, followed by horses, sheep, goats, and pigs. Emissions from lactating dairy cows are particularly high. Farming produced a half of all methane emissions in the UK in 2014. By contrast, farming accounts for barely 1% of total UK carbon dioxide emissions – the most well-known greenhouse gas – mostly from energy used on farms for fuel and heating; substantially more is produced in the energy intensive process of fertiliser production.

The UK's emissions overall are expected to fall somewhat by 2020 as measures are implemented across all sources to get the UK's contributions down significantly in order to fulfill international commitments. Yet the vast majority of the UK's farmers don't seem to be taking the issue seriously. Defra's 2016 Farm Practice Survey found that only 9% of farmers believed that it was 'very important' to consider greenhouse gas emissions when making decisions relating to their land, crops, and livestock, and a further 39% thought it just 'fairly important'.[x] The dairy industry, however, reckons that 78% of dairy farmers are currently taking action to reduce greenhouse gas emissions on their farms, the highest percentage out of all farming sectors.[xi]

Nitrous oxide is a particular concern since it is between 200 and 300 times more potent as a greenhouse gas than carbon dioxide. Together with methane (23 times more potent than carbon dioxide), these two gases have a much more substantial effect on atmospheric temperature. When the amount of nitrogen-rich fertiliser (either synthetic fertiliser or manure applied or excreted onto land) added to a soil is more than can be taken up and used by plants, bursts of nitrous oxide are often produced by soil bacteria that make use of the spare nitrogen as an energy source. Limiting inputs of nitrogen to just the amount likely to be usable by plants can reduce emissions. This is easy in principle, but not necessarily in practice. Using the latest nitrogen-efficient varieties of crops and including nitrogen-fixing crops and soil cover crops in farm rotations helps too. However, nitrous oxide from farm sources in the UK isn't falling very much, unlike emissions from other sources. Between 1990 and 2014, nitrous oxide emissions from industry declined by at least 80%; emissions from agriculture fell by 15% at most.[x]

One way of reducing the 'raspberry tart' output of ruminant livestock is, of course, to have fewer of them. In terms of long-term sustainability, it's more efficient for us to consume the products of crops directly – wheat and maize for instance – rather than feed such crops to livestock and for us to eat them instead. In 1875 there were about six million cattle on UK farms. The population rose steadily to a maximum of 15.2 million in 1974 then gradually fell back (sometimes more steeply such as during the BSE crisis) to about 10 million in 2005 and has stayed roughly at that number since.[xii] This reduction in numbers undoubtedly reduced methane outputs at the time but not in recent years. With questions being raised about the impact of overconsumption of red meat on human health, maybe livestock numbers will fall further in future. Worldwide, because more affluent people in some of the better-off developing countries are eating increasing amounts of farmed meat, cattle numbers are increasing.

LIVING IN A GREENHOUSE

The primary greenhouse gases in Earth's atmosphere are water vapour, carbon dioxide, methane, nitrous oxide, and ozone. Without greenhouse gases, the average temperature of the Earth's surface would be about −18 °C rather than the present average of 15 °C so they are essential to insulate us. But human activities since the beginning of the Industrial Revolution have produced a 40% increase in the atmospheric concentration of carbon dioxide, mostly from the burning of carbon-based fuels (principally coal, oil, and natural gas), along with deforestation, soil erosion, and animal agriculture. Our planet is overheating. Recent estimates suggest that on the current emissions trajectory, the Earth could pass a threshold of a 2 °C increase in global warming, which the United Nations' Intergovernmental Panel on Climate Change designated as the upper limit for 'dangerous' global warming, by 2036.

Making sure that cattle are not fed an excess of supplements in their diet and installing anaerobic digestion units to treat manure from larger dairy cattle farms can help to reduce farm methane. There is also evidence that feeding cows a diet supplemented with linseed oil and calcium nitrate decreases significantly the amount of methane they produce.[xiii] Nevertheless, while methane emissions in the UK from non-agricultural sources declined by 60% between 1990 and 2014, methane emissions from agriculture barely fell.[xiv]

With extreme weather events in many parts of the world – prolonged droughts, wildfires, unusually powerful tornados, and devastating floods – responsible for numerous deaths, and the likelihood that low-lying coastal areas, even whole nations, will disappear as sea levels rise, climate warming's implications for farming are often overlooked. Here in the UK, with predictions of increasing rainfall and stormy conditions in the north and west of the country – but more summertime droughts in the south – farming is going to be considerably affected by the changes it's bringing.

A warming climate for southern Britain is likely to produce more summertime droughts and an increased frequency of wintertime flooding from heavy rain; in consequence, how farmers manage their water supplies and what crops they grow are going to be increasingly vital. Extracting less water from underground aquifers and watercourses will limit pressure on water and wetland habitats and the species they support. Furthermore, using less water generally and saving more will lower production costs, improve profit margins, and make a farming business more resilient longer term.

A warmer climate and rising atmospheric carbon dioxide levels have the potential to increase crop yields. Crops more suited to dry conditions such as durum wheat, grain maize, varieties of legumes including peas and clovers, soybeans, sunflowers, millet, sorghum, chickpeas, lucerne as a substitute for ryegrass pasture, quinoa, and other oilseed and starch crops could become more commonly planted. Sunflowers attract several species of bees and hoverflies to pollinate them though most of the other drought-resistant crops bring no particular wildlife advantages.

Herbal leys incorporating plants such as chicory, Sainfoin, and plantain on the other hand – more drought resistant than ryegrass-clover pasture – could become more commonplace for lowland sheep and cattle grazing, boosting invertebrate populations that are associated with a more diverse plant species mix (see Chapter 6). Herbal leys have another big advantage too: they require very little or no fertiliser, reducing the demand for a synthetic product that generates large volumes of carbon dioxide in its manufacture.

Many crop experts see GM technology coming to the aid of crops facing drought conditions. For instance, barley naturally has a gene responsible for opening and closing tiny pores on the surface of its leaves; pores that are used by the plant for gas exchange but which also provide an exit route for water vapour. They tend to close at night to conserve water and open in the day to allow photosynthesis. By putting the naturally-present gene controlling this process into overdrive, barley plants have been developed with pores which close more readily when water is scarce, retaining the water content of the plant and making them more resilient in droughts. There was no 'trade-off' associated with this change because the plants grew as well in water-abundant soils as their non-modified counterparts. Undoubtedly, GM drought-resistant varieties of crops such as maize and oilseed rape will be promoted by scientists and seed companies, but whether they will gain public acceptance in the near future is debatable at best.

Reducing the impact of climate warming is so important that UK farmers are going to have to play a much more positive role in reducing their emissions of greenhouse gases (and of other pollutants from farms) in the future. Appropriate measures will need to be incorporated into whatever agri-environment schemes the four agriculture departments in the UK adopt in future and their implementation enforced. Considering that farmers, both in our uplands and lowlands, are on the frontline of the probable implications of climate warming and are going to need to adapt to it more readily than many other occupations, the industry needs to embrace a more positive response than it perhaps has hitherto.

———————————

Clearly today's farmers have to understand much more than livestock husbandry or traditional crop management. They need to be familiar with environmental regulations covering wildlife and pollution, animal welfare issues, the safe use of hazardous substances, equipment maintenance, the likely impacts of climate warming, and much more. Whilst the majority of younger farmers have obtained farming qualifications, many older farmers never did. Farming in the UK has an aging workforce: around 30% are over 65 years old while the proportion of farmers below the age of 35 is just 3%. What's more, while the number of farmers in the 35–44 age category is decreasing, the proportion in the oldest band, 65 years and over, is increasing.[ix]

More than one farmer I interviewed in the course of writing this book suggested that farmers should undertake continuing professional development (CPD) and that regular attendance on appropriate college courses should be linked to receiving their subsidy payments! The suggestion opens up the possibility of incorporating components in such CPD courses about the environmental consequences of farm operations, wildlife conservation, and habitat management for wildlife.

CPD describes the learning activities professionals engage in to develop and enhance their abilities throughout their careers. Teachers, nurses, and others are obliged to do it. Many other professions such as plumbers and gas installers have to participate in accredited college courses as regulations change and to increase their qualifications. For example, under the Code of Professional Standards that all members of the Chartered Institute of Plumbing and Heating Engineering (CIPHE) agree to abide by, members are required to participate in continuous professional

Gases galore. Cattle are a major source of methane, a greenhouse gas 23 times more potent than carbon dioxide.

development; CIPHE currently recommends that members should obtain 30 hours of CPD on an annual basis.[xv] But farmers are under no obligation to do any of this, even as further complex and binding taxpayer-subsidised regulations are implemented.

Some farmers, perhaps most, will be concerned if they have yet more obligations to fulfil; after all, many who voted to leave the EU (and the CAP that has protected their interests for decades) apparently believe that the governments of the UK will insist on *less* bureaucracy than the EU. That seems, at best, a forlorn hope. One livestock farmer I met showed me boxes full of papers relating to regulations on his farm: the details of an agri-environment agreement field by field; guidance on claiming his BPS and greening payments; the implications of the Nitrate Vulnerable Zone covering part of his farm; and more. 'I think I knew about all this at some stage, but I don't have a bloody clue what some of it means now,' he admitted despairingly. 'If I get an inspection, I'm not sure all these papers are even in order. I'll just have to accept whatever fine they impose!'

CPD for farmers has its supporters. Not surprisingly, one is Lantra (until 1996, the Agricultural Training Board) whose primary purpose is to develop qualifications and training courses delivered by training providers across the UK. Groundcare, falconry, chainsaw maintenance, beekeeping, dry stone walling, all-terrain vehicle use, quad and tractor driving, and rural building conservation are just some of the areas it specialises in already.

'CPD is vital for farmers,' says Sallyann Baldry, Lantra's Business Development Director. 'We operate a number of registers to help people track and record their CPD; we have one for poultry-meat producers in partnership with British Poultry Training which has 5,200 users and we are just about to produce a similar one for egg production. We recently launched "Pest Passport" which enables operators to record their compliance with new legislation around the safe use of aluminium phosphates and rodenticides. To introduce it for all farmers, in our experience training must be made mandatory to be effective. It's unlikely that farmers would pay unless it's a legislative requirement or their customers (i.e. the better supermarkets) make it a contractual obligation. Most Lantra training is short, one to five day courses which are of a practical nature. As a point of principal Lantra is keen to improve the environmental credentials of our courses and as a charity we see the issue of skills development in its widest context, in particular with the aim of improving/conserving the environment, its flora and fauna as integral to our mission.

'We also know that many farmers prefer to have the training at their farm and will call upon private training providers, many of them working outside the term times that colleges stick to. We have a network of nearly 400 freelance trainers who can deliver on-farm if required. It avoids farmers travelling away from their workplace, the equipment and context of the training is site specific, and it lowers the risk of the trainee finding it hard to relate the learning to their actual place of work,' she adds.

City & Guilds, another leader in skills development that delivers services to training providers, employers, and trainees across a variety of sectors, already offers a range of agriculture and conservation-related qualifications and services to provide rural managers with skills and knowledge to support their management of land and wildlife. 'At present, there are differing views within the farming sector with respect to compulsory training or engagement with CPD, although it's widely recognised that CPD is important if the farming sector is to be competitive and embrace new technologies and working practices. And we'd be happy to work collaboratively with the agriculture and related industries to support them on any skills initiative to ensure that the best possible management practices are promoted,' comments Dr Robin Jackson, City & Guilds' Senior Manager for Land Based Industries.

But Chris Moody, Chief Executive of Landex, '*Land Based Colleges Aspiring to Excellence*', *which has* 41 member colleges and universities in the UK, is concerned about CPD being made compulsory for farmers. 'Compulsory CPD for farmers is a very contentious issue; it's not the CPD itself but the compulsory element that is more difficult. Many would regard it as yet more farming regulation and/or view it as the first step towards the introduction of a kind of licence to farm,' he says.

Another issue is that not all agriculture courses that are run in colleges and universities for young people hoping to enter the industry include aspects of wildlife conservation as mandatory elements. With farming having such a huge impact on wildlife habitats and species, that's more than a little surprising. It's certainly something that needs correcting.

'Such courses are not obligatory for all agriculture students,' remarks Mr Moody. 'The content and examination of long, publicly funded qualifications is determined by the Awarding Bodies' – of which there are at least 120 in the UK including Lantra and City & Guilds – 'not as is commonly imagined by the colleges themselves. The Awarding Bodies are in turn required to consult employers on content, and to have their qualification content approved by the relevant government department. If providers had the autonomy to develop and approve, in consultation with employers, the content of the courses that they deliver would be significantly different.'

Along with CPD, it's an issue that needs tackling if our farmers are expected to better understand the wildlife and environmental consequences of what they do, and they continue to receive public support post Brexit to produce food *and* wildlife gains in our farmed countryside.

––––––––––––––

There is another issue which is already impacting on farming in the UK and which is seemingly gathering momentum: the reintroduction of animal species that once made our forests, hills, and valleys their home but which were hunted to extinction here, some of them in the distant past. They are species such as the European Beaver, already re- establishing on one or two rivers; White-tailed Eagles now well re-established on parts of the north-west coast of Scotland but not elsewhere; and the European Lynx. Other conservationists want to do much more than this: rewilding is the idea that significant tracts of countryside should be taken out of exploitation for farming or forestry (or any other human industry), put over completely to wildlife conservation and controlled tourism, and missing species known to have inhabited the area reintroduced. In a country such as the UK with a crowded human population, it's only in the more remote parts of Scotland and maybe Central Wales that rewilding on a significant scale might be practicable.

The charity Rewilding Britain has been set up 'To see at least one million hectares in Britain supporting natural ecological processes for the benefit of people and nature'. Somewhat oddly, the 'species of interest' it would like to see re-established – or their existing restricted ranges expanded – are all animals,[xvi] and they include many, such as fish and whales, that would not impinge on farming. Others such as Night Herons and White Storks would have no obvious implications for farmers either except that storks often feed on wet grasslands. But many others would. Most farmers baulk at mention of European Lynx or Grey Wolf, neither of which have yet been returned to the UK. Both are, of course, powerful predators that can kill deer and are likely to make an occasional meal of smaller farm livestock too. With farming occupying over 70% of the

UK's land area, with a fair proportion of that being sheep-grazed hills, moors, and mountains rather than intensively-farmed lowlands, predatory mammals are going to come into pretty intimate contact with farm livestock from time to time. Ideas about wolf reintroductions might seem outlandish yet bringing back European Lynx – especially useful for reducing burgeoning Red Deer population in Scotland where they do considerable damage to woodlands and plantations – is a more likely possibility.

ON THE WILD SIDE

The word 'rewilding' was coined by US environmentalist and activist Dave Foreman, one of the founders of the group Earth First!, who went on to help establish the Rewilding Institute in the US. It was promoted as a method to preserve ecosystems and reduce wildlife losses. In 1967 Robert H. MacArthur, a Canadian ecologist, and Edward O. Wilson, a leading US ecologist, established the importance of considering the size and isolation of wildlife conservation areas, stating that protected areas remained vulnerable to extinctions if small and isolated. While the rewilding concept is most often applied to largely uninhabited land areas and 'wilderness', it has also more recently been applied in several high human-density European countries including The Netherlands. Here, the Oostvaardersplassen nature reserve is sometimes considered a rewilding project on just 56 square kilometres of wet grassland, reedbed, and woodland on polder reclaimed from the sea in 1968. Increasingly the term has been purloined to refer to many species reintroduction projects – European Beavers, Common Cranes, Wild Boar, and others – albeit these require appropriate areas of restored or recreated habitat that is not necessarily uninhabited by people.

White-tailed Eagles, renowned for their Osprey-like ability to seize fish from lakes and inshore seas, were reintroduced in the 1970s to parts of north-west Scotland, where they have added lambs to their diet. How many of these lambs are killed rather than taken after they have died of other causes, and how many might have died anyway in what are rather testing environments for livestock, is somewhat unclear. What is much clearer is that these massive and impressive eagles attract tourists who bolster otherwise-struggling rural economies. There is acceptance that the eagles are here to stay and conservationists, working with local sheep farmers, have agreed a package of measures including compensation for lamb losses and displacement eagle feeding locations to draw these magnificent hunters away from prime lambing areas.

Nevertheless, attempts to reinstate White-tailed Eagles to other parts of the UK have failed, notably in Suffolk. The RSPB and Natural England had argued that the large areas of coastal marshes and waterways, many protected as reserves, could easily support them with fish, rabbits, and carrion. But Suffolk's sheep farmers and their representatives were seriously concerned that lambs might well be eyed up by hunting eagles. Soon, pig and free-range poultry farmers joined the fray. Public opinion started to turn anti-eagle and the proposal was dropped.

So it is that large predators are not going to be greeted favourably by livestock farmers even when compensation schemes – like those in some other European countries – are mentioned, and estimates of tourism income are dangled in front of rural communities. The Lynx UK Trust, a group of conservationists and others dedicated to reintroducing the European Lynx into the ecosystem of the British Isles, has been considering trials of these attractive hunters in Norfolk, Cumbria, and Aberdeenshire.[xvii] Now though, it is focussing on Kielder Forest (an extensive area of conifer plantations) in Northumberland, which has a large Roe Deer population, the main prey for lynx. Having carried out a lengthy consultation with landowners and farmers in the area, in July 2017 the Lynx UK Trust submitted an application to Natural England for permission to carry out a five year trial reintroduction with six Eurasian lynx (four females and two males) captured in the wilds of Sweden.

Hunted to extinction in the UK over a millennium ago, where lynx occur on the European continent they rarely attack farm livestock and, being nocturnal, are not often seen. Despite this, sheep farmers in particular remain concerned by the proposal and the possible future implications if the cats eventually are reintroduced more widely or escape from trial areas.

Cute? Most farmers don't think so but many conservationists want to see Eurasian Lynx returned to Britain (courtesy of Bernard Landgraf).

Farmers' vociferous objections sometimes drown out the often more positive response to such proposals from small community-based businesses, local shops, pubs, restaurants, B&Bs, and hotels, many of which are struggling to survive in some rural parts of the UK. Several such local businesses can cash in on the tourism revenue that often results from these schemes, especially if they involve animals that can be seen in daylight. Farmers need to remember that they receive substantial subsidies from taxpayers to keep farming; no other rural business gets that kind of support to keep them viable!

On the 9,000 hectare Alladale Estate in Sutherland the landowner, Paul Lister, is into rewilding in a big way. Planting thousands of native trees and protecting them from deer grazing, he intends to return the land to something akin to what it would have been naturally a couple of thousand years ago. Vast areas of the Highlands consist of mile after mile of heather and bracken because of long-term livestock grazing, deer browsing, and vegetation burning, which put paid to most of its natural tree cover. But Mr Lister isn't stopping there; he wants to bring back Brown Bears and Grey Wolves to his estate and to other neighbouring land subject to it being fenced to prevent the animals from getting out. Fearing that these predators will escape when deep winter snow provides a means of vaulting the fences, landowners and farmers nearby are concerned about the impact of the wolves in particular. Whether he gets licences to do these reintroductions remains to be seen.

European Beavers are already living wild in part of Argyll where a trial was done over several years to assess the implications of their presence. There are also beavers in the wild in parts of Devon and on the River Tay, the result of escapes or illegal reintroductions. In November 2016 Roseanna Cunningham, the Scottish Government's Cabinet Secretary for Environment, Climate Change and Land Reform, announced legal protection for the Tay and Argyll beavers but stated that the species would still need to be 'actively managed',[xviii] whatever that means. 'Beavers promote biodiversity by creating new ponds and wetlands which in turn provide valuable habitats for a wide range of other species. We want to realise these biodiversity benefits while limiting adverse impacts on farmers and other land users. This will require careful management', she said.

Decisions on reintroducing and giving legal protection to beavers in other parts of the UK would be taken by the relevant devolved governments and their conservation agencies, although public opinion is generally in favour. Some farmers are concerned that beavers damage riverside banks and cause flooding, but experience in other European countries doesn't suggest a significant problem.

As many farmers across Europe have found, Britain's farmers might well discover themselves having to co-exist with some mammals and large birds of prey that have not inhabited our countryside for a very long time; this absence of top predators is entirely unnatural. The tide of public opinion appears to be moving strongly in that direction and increasing calls from conservation organisations are not going to subdue.

Endnotes

i   'Q and A about Genetically Modified Crops,' International Service for the Acquisition of Agri-biotech Applications, September 2016.

ii   'Recent trends in GE adoption,' US Department ofAgriculture Economic Research Service, 2016.

iii 'Our policy on GM organisms,' Defra Policy paper, 2010 to 2015: Government policy: food and farming industry, updated 2015.

iv   'How are GM crops regulated?' The Royal Society, 2016.

v   http://www.foeeurope.org/sites/default/files/news/foee_background_maize1507_short.pdf.

vi   'Use and Impact of Bt Maize,' R. L. & K. A. Hellmich, 2012. *Nature Education Knowledge* 3(10):4.

vii 'Genetically modified crops 2016,' www.greenfacts.org/en/gmo/index.htm.

viii 'Cultivating The Future: How can 20 years of GM debate inform UK farm policy?' Agricultural Biotech nology Council, 2016.

ix   '2010 to 2015 Government policy: greenhouse gas emissions,' Defra, 2015.

x   'Farm practices survey February 2016 - greenhouse gas mitigation practices,' Defra, 2016.

xi   'The Dairy Roadmap 2015,' AHDB Dairy, Dairy UK, the National Farmers Union, and others.

xii Agriculture: Historical Statistics. House of Commons Briefing paper number 03339, 21 January, 2016.

xiii 'Additive methane-mitigating effect between linseed oil and nitrate fed to cattle,' J Guyader et al, 2015. *American Society of Animal Science* 93:3564–3577.

xiv 'Agriculture in the UK 2015,' Defra and Devolved Governments, 2016.

xv   www.ciphe.org.uk/professional-members/cpd/.

xvi www.rewildingbritain.org.uk/rewilding/reintroductions/.

xvii http://www.lynxuk.org/publications/EngLynxConsult.pdf.

xviii https://news.gov.scot/news/beavers-to-remain-in-scotland.

CHAPTER 14

THE NEW FURROW

Since 1973 the EU's Common Agriculture Policy (CAP) has determined the direction and support of UK farming. A 20-year-old farmer setting out in that year would have been 63 years old by the time the decision was taken in 2016 to withdraw from the EU. He will at least be three years older still before the UK Government and the devolved governments of Northern Ireland, Wales, and Scotland will be free of the CAP and able to regulate Britain's farming. For a large number of our farmers, this is novel territory; they will not have known a farming era governed other than by the CAP.

While it has been a long-held belief by successive UK governments that farmed land should benefit wildlife alongside producing food, the constraints of the CAP and the necessity to agree farming policy across 28 member states have limited what could be achieved. For much of its existence, the CAP has been responsible for promoting farming policies which have destroyed huge amounts of farmland wildlife habitat, mostly in the lowlands but in more subtle ways in our hills and uplands too. Only in recent years has the CAP allocated significant sums to improving farm wildlife and other heritage features, trying to put back a little of what has been lost over decades and care better for what remains. The CAP money allocated to farming remains heavily loaded in favour of the Basic Payment Scheme (BPS), the subsidies for receipt of which farmers are required to do very little to conserve wildlife on their farms. Many upland farmers have become totally reliant on subsidies, reducing any incentive to explore other land management options or diversify their businesses.

Brexit has created an opportunity to reform farming substantially, to plough a new furrow as it were. The majority of farmers, it seems, voted in favour of leaving the EU in spite of the enormous cash supports the CAP has provided them with since its inception. They appear to have gambled that leaving the EU would somehow unshackle them from the CAP's regulations and reduce the bureaucracy they endure. What they did not necessarily realise is that if the UK is to continue to trade with EU countries after we leave, we will almost certainly have to adhere to whatever farming standards the EU sets anyway! It's hardly likely that the four governments within the UK that will be entirely responsible for farming policy in their own countries are going to short-change on bureaucracy. What is understandably considered as bureaucracy by most farmers is in civil servants' books the essential administration and checking that's required if taxpayers' money is going to be dispensed.

Post CAP, while the UK and devolved governments' ministers issue generalised statements about 'continuing to support Britain's farmers', for the first time in four decades the budgets allocated to farming will have to compete directly, and very visibly, with other calls on Exchequer funds, not least the NHS and education. As a result, many commentators think that the BPS payments farmers receive under the CAP will cease altogether or be reduced in stages over a period of just a few years until they are eventually removed. Professor David Harvey of Newcastle University, a leading agricultural economist, has proposed paying farmers a lump sum after we leave the EU, a kind of 'exit bond' indexed to their farm area and the equivalent of the capitalisation of such a phased elimination.[i] However, would what was perceived as a 'golden handshake for doing nothing' be acceptable both publicly and politically in these difficult financial times when most people, at least in the public sector, have received almost no pay rise for several years?

New Furrow 1: A per hectare subsidy, currently the BPS, should continue to be paid to hill and mountain farms (above a land elevation limit to be defined) but capped, on a declining scale based on land area, at no more than £200,000 per farm per annum. Lowland farms would receive a continuing subsidy payment (immediately capped at £200,000 per farm) on a declining scale to zero for a maximum of five years after the UK leaves the EU.

As the BPS is often the main income for the smaller and most isolated hill and mountain farms, removing it completely would see many go out of business. On a limited scale this loss would probably not be a significant problem in wildlife terms; it could mean better development of the vegetation that is free of grazing and intermittent burning. However, on a larger scale, removing sheep and cattle grazing entirely from large parts of our uplands could slowly but surely have a detrimental effect on much wildlife habitat – heather moors and blanket bogs for instance – where light livestock grazing is essential to maintain their wildlife value accumulated over centuries, even millennia. Some grazing on some upland habitats is essential to retain their wildlife interest, although heavy grazing has damaged much of our upland and mountain habitats and should be substantially reduced. Also, large areas of upland without hill farmers could result in sheep ranching or more conifer plantations, neither of which provide positive wildlife gains.

Fixing a maximum BPS payment for hill and mountain farms as is done for all farms in Wales, where the payments are capped currently at €300,000 – about £274,000 per farm – would be a significant saving of public money. The Welsh limit might well be seen as too generous, though payments there reduce on a sliding scale with farm size, and a lower cap might be more realistic. As discussed later, the cash saved could then be directed into agri-environment schemes to improve existing farmland wildlife habitats and create new ones.

But it's difficult in profitability terms to justify continuing to pay subsidies to lowland farms that need it least. Most don't need any public support; they are businesses having to take the risks that many other businesses take day to day, and should be able to stand on their own. A period of adjustment might be necessary to allow such farms to do any re-organising of their businesses, maybe five years, but all lowland farm subsidies should cease.

New Furrow 2: Agri-environment schemes should remain the primary mechanism for creating and managing farmland wildlife habitats. Such schemes should be reviewed:
a. to examine whether more flexibility can be built into their prescriptions;
b. to see how they can be made more outcome-based; and
c. to determine whether income foregone payments should be re-assessed because of recent research indicating that certain prescriptions result in crop yield gains not losses

Agri-environment schemes were first introduced into EU agricultural policy during the late 1980s as an option for member states. Since 1992, they have been compulsory for member states but remain optional for farmers. They are the only significant mechanism to encourage farmers to conduct their operations in a more wildlife-friendly way; they provide farmers with an incentive for creating farm wildlife habitats and for managing any habitats that remain on a farm. And they are as described: 'agri-environment', farming in a way that produces both food and wildlife! They are not a non-profit alternative to farming. Their practical benefits have been discussed in Chapters 5, 6, 7, and 9 in particular.

If the British taxpayer wants to know that our farmed countryside – which occupies 70% of our land – not only produces food but supports a wealth of wildlife too, agri-environment schemes offer the potential way forward. They can also be used to ensure that farming's impact on water supplies and water quality is minimised; that its contribution to climate warming is reduced; and that it provides, especially in the uplands, natural 'sponges' to slow flood waters. These are all part of what the Country Land and Business Association (CLA) refers to as 'ecosystem services' that the government can purchase from farmers.

Agri-environment schemes offer payments for work done by the farmer – creating a pond, a copse of trees, and hedge planting for example – but many of the payments are based on profit foregone if, for instance, part of the farm which would normally be used for crop growing is converted to flower-rich margins. The payments per hectare can be significant and compensate the farmer accordingly. They have had considerable local success, such as the restoration of habitat and feeding resources for Cirl Buntings in Devon (see Chapter 7). What they have not yet demonstrated is whether they are slowing or stopping the seemingly inexorable decline countrywide of many once common farmland birds, butterflies, and other invertebrates.

Recent research doesn't suggest that there is always a cost to the farmer in establishing habitats for pollinators or other wildlife using agri-environment supports; some such prescriptions might actually increase crop yield.[ii] A study on a 900 hectare commercial arable farm growing cereals, oilseed rape, and beans in central England found that creating wildlife habitat in lower-yielding crop edge areas led to increased yield in the cropped fields, an effect that became more pronounced over six years.[iii] This is presumably because of larger overall insect pollinator populations and enhanced populations of predatory insects such as ground beetles, which can reduce insect disease infestations (of aphids for example) in the crops. In fields without flower-rich margins, crop yields were reduced by up to a third at the field edge, probably because of soil compaction, competition for light and water resources with adjacent hedges and trees, and increased pressure from pests and weed species.

Since the first such schemes were launched, they have become more targeted at particular farmland habitats and species and considerably more detailed in their content and in their prescriptive nature. Guidance notes or manuals issued with the schemes – which are different in each UK country (though they have many options in common) – run to hundreds of pages. Many farmers and some commentators consider that they have become overly prescriptive and inflexible, concentrating on the instructions more than the intended outcomes and not taking enough account of differing farming circumstances and changing year-to-year conditions. Their design and implementation needs to better reflect the natural fluctuations of seasonality and the farming calendar.

The UK's leading conservation organisations – which together have a membership of over six million people – have called for all farming subsidies to be abolished and replaced with payments for agri-environment measures to deliver substantial wildlife and other environmental gains along with food. Professor Alan Matthews, Professor Emeritus of European Agricultural Policy at Trinity College, Dublin, doesn't support the idea of making agri-environment schemes compulsory and would prefer to have them made a requirement if a farmer is receiving any government aid such as investment grants or cash incentives to encourage young farmers into the industry.

Food and wildlife co-existing. Barley for food; a bird seed mix for invertebrates, small mammals, and birds. Newlands Farm, Dorset.

The problem with such an approach is that take-up of agri-environment schemes might then continue to be pepper-potted around the countryside, diminishing the chances of meaningful progress in getting wildlife habitats and their species reinstated on any large scale. Essential habitat connectivity from farm to farm to allow poorly mobile species to move in response to climate warming will be limited, and it will not provide the means of addressing the enormous declines in farm wildlife we have witnessed over decades countrywide.

New Furrow 3: Cash saved by removing and capping BPS subsidies to farmers should be redirected into agri-environment schemes so that they are able to have a significant impact in recreating wildlife habitats and better managing existing areas of wildlife habitat on farms. Emphasis needs to be placed on connecting up habitats farm by farm and on landscape scale gains in wildlife.

More cash in agri-environment budgets would provide an unprecedented opportunity to make good progress in returning wildlife to farmland. In 2015, the UK was allocated a budget of €3.1 billion for BPS payments (subsidies) but just €600 million for agri-environment payments to help restore farmland wildlife. If the UK's farmland is to offer significant space for wildlife alongside producing food, that enormous budget difference has to be addressed. Eliminating BPS payments for all lowland farmers and capping them for hill and upland farmers would achieve that redistribution if the BPS savings were reinvested in agri-environment measures.

In 2016, the total area of land in higher-level or targeted agri-environment agreements in the UK was just under 2.4 million hectares; 1.4 million hectares in England; 0.7 million hectares in Scotland; 0.2 million hectares in Wales; and 0.1 million hectares in Northern Ireland.[iv] While the UK total increased annually through the 1990s until about 2012, since then it has fallen by a million hectares. The greatest decline in agri-environment land has been in Northern Ireland followed by Scotland and less so in Wales. In England the area has stayed about the same since 2014.

Official audits of some agri-environment schemes, while generally very supportive of what the schemes are achieving, have identified flaws. For instance, the Auditor General for Wales published an audit in 2014 of Glastir, the Welsh agri-environment scheme commenting that participation in the scheme was well below the Welsh Government's targets, some of which were unrealistic, and that measures to evaluate the scheme's success were yet to be developed.[v] It found that at the application stage, the Welsh Government did not collect any information on existing farm management practices to give it assurance that the landholder needed to make changes on joining the scheme. Moreover, poor existing land management practices by some farmers (such as allowing pollution of watercourses) were not tackled before a Glastir scheme was agreed. In other words, it was difficult to know how much additional environmental value had been achieved for the money spent.

New Furrow 4: All farmers should be required to produce an agri-environment plan (AEP) for their farm to be agreed by the respective government or agency in each UK country as the basis of their agri-environment agreement. Conservation NGOs and the private sector should be able to charge on an agreed sliding scale for developing such plans with farmers. AEPs should contain mandatory requirements for a set of minimum environmental, animal welfare, and safety standards (current so-called 'cross-compliance' requirements). For hill and mountain farmers the emphasis in an AEP must be on reducing livestock numbers overall and fencing out hill slopes, especially near streams, to eliminate livestock grazing in order to encourage more vegetation development for wildlife and to help reduce the devastating impact of downstream urban flooding. For lowland farms the emphasis should be on the appropriate management of existing, and the creation of new, wildlife habitats.

A fundamental difficulty with agri-environment schemes as currently required by the CAP is that they are not compulsory for the farmer! Consequently, there has been a tendency for such schemes to 'pepper pot' the farmed countryside, limiting the gains associated with joining up wildlife-rich zones from one farm to another, for instance along the catchment of a river or by connecting up surviving flower and insect-rich downland fragments. This has been reduced – but not eliminated – by those administering the schemes in some parts of the UK encouraging applications from within particularly important targeted land areas. Nevertheless, only if farmers have an innate interest in managing their land in a more wildlife-sensitive way, and/or are attracted by the payments, are they likely to discuss entering a scheme and making an application.

In our hills and uplands the main agents of change have historically been over-grazing by too many livestock (see Chapter 9) – sheep especially – and too much vegetation burning. This has resulted in many upland and mountain areas having closely-grazed turf consisting of very few plant species (and thereby far fewer animal species), albeit making access easier for hikers. It also causes potentially devastating floods in urban areas downstream by enabling heavy rain to flow off such slopes into watercourses much more rapidly than it would if the vegetation was better developed and able to slow some of the water. Concentrating future upland agri-environment schemes on fencing out livestock from hill slopes, especially around streams, to allow the natural vegetation to recover and mature, and lowering livestock numbers overall in the uplands, would help return more species to our upland and mountain regions and reduce the damaging impact of flooding.

New Furrow 5: The existing greening payments added to BPS (subsidy) payments should be abolished; the more effective elements of greening, especially measures designed to reduce soil and nutrient runoff, should be incorporated into basic compliance requirements in AEPs for all farmers. The cash saved should be redirected into agri-environment schemes.

Farmers are paid an additional 'greening payment' on top of their BPS subsidies provided they undertake agricultural practices that are beneficial for the climate and the environment. These 'greening' payments add about a third more cash to the BPS payment but are a dog's breakfast of measures that do little to improve the lot of wildlife on any farm. Three elements make up the greening requirements: crop diversification, Ecological Focus Areas (EFA), and permanent grassland. The rules for implementing them are fiendishly complicated (see Chapter 4). They account for a considerable expenditure of CAP funds; although they deliver some benefits in terms of reducing runoff and nutrient losses from fields, they deliver

*Natural sponge. Lightly sheep-grazed blanket bog soaks up
rain and reduces flood peaks downstream. Migneint, Snowdonia.*

very minor wildlife gains. The elements of greening that are of at least minor benefit could be incorporated in AEPs developed for every farm.

New Furrow 6: All farmers receiving subsidy and agri-environment payments post CAP should be required to attend college-based courses for continuing professional development (CPD), the content of which must include farm wildlife habitat management, environmental protection, and conservation. Moreover, all college agriculture courses should contain compulsory units on farmland wildlife conservation and sustainable land management.

During discussions with farmers for writing this book, more than one questioned why it was not compulsory for farmers in receipt of subsidies to have to attend regular college-based courses for continuing professional development (CPD). It seems a perfectly reasonable suggestion: it would bring farmers 'into line' with many other professions where CPD is considered essential; and it would increase the standard of farming knowhow across an industry in which many older farmers have never participated in any formal training. Providers of such farming CPD would need to include conservation courses as compulsory components and attendance would be required for all farmers in receipt of any continuing BPS payments in the uplands or agri-environment payments. Agriculture courses for students at all levels should likewise contain compulsory units on wildlife conservation and environmental protection.

New Furrow 7: The wildlife advantages of organic farming and sustainable farming initiatives that require high environmental standards on farms need to be promoted more actively by governments and government agencies. This would encourage more farmers to adopt such standards and/ or convert to organic production using the supports currently available. The financial supports to convert to organic from non-organic farming should be independently reviewed to examine whether they should be modified to get more farmers in different farming sectors to convert to organic production. At the same time, government support across the UK for organic food needs to be bolstered with marketing campaigns and by encouraging home-grown organic produce to be used in schools and by public sector organisations.

Organic farming has proven benefits for farmland wildlife and specific agri-environment measures are available to help farmers fund the process of conversion from non-organic to organic production. Together with more sustainable forms of non-organic farming such as that promoted by LEAF (Linking Environment and Farming), it is clearly advantageous that support for these more environmentally friendly and more wildlife-sensitive farming systems should be bolstered.

Governments in the UK need to do more. Steven Jacobs, Business Development Manager for Organic Farmers and Growers, refers to the way the Danish Government actively promotes organic production by supporting growers, marketing effectively, encouraging retailers to stock wide choices of organic produce, and urging wholesalers and caterers for the public sector to stock organic products (Chapter 10).

New Furrow 8: The farming industry should promote and implement improved land management measures to better safeguard agricultural soil and improve its organic matter content. Enhanced slurry management and less soil compaction and runoff to reduce pollution of waterways and greenhouse gas production will also require better land management. This encompasses the use of cover crops and replacing ploughing with 'no till' wherever possible. Such measures should be included as part of the basic element in all AEPs.

An unusual sight. Chris Davies, Cwmachau Farm near Brecon haymaking.

Concern is mounting about the state of farmland soils, particularly in the lowlands. Decreased levels of soil organic matter as well as compaction and surface water runoff carrying nutrients to watercourses are wasting precious resources that all farming ultimately depends upon, depleting the soil biota which is essential for nutrient recycling. Some existing measures like planting soil cover crops to provide ground cover after main crops have been harvested will undoubtedly help reduce soil runoff and wind erosion of bare soils. But the UK's four agriculture departments and the farming industry itself need to consider more actively supporting modified land management such as 'no till' to replace ploughing more widely in order to improve soil organic content and its biota and reduce compaction and runoff. Better slurry management and application techniques are also essential. Leaving such an important issue to voluntary implementation might not be sufficient; soil cover crops and improved slurry application should be a compulsory component of all AEPs for farms.

New Furrow 9: The farming industry should promote reduced, careful, and targeted pesticide use. Prophylactic use of pesticides should be strongly discouraged and the precautionary principle invoked to ban the use of any pesticide(s) where objective evidence suggests that their use is linked to significant wildlife declines.

The EU is currently reviewing pesticide legislation which is likely to result in fewer pesticides being licensed for farm use. Neonicotinoid insecticides are presently under a temporary ban because of evidence that they might be contributing to bee declines and the European Commission is proposing a complete ban on their use. Because of greater scrutiny and concern, other pesticides could be banned or their use restricted too.

What pesticides the EU decides to ban or restrict is likely to have ongoing implications for the UK's farmers given the probable need to continue to follow EU rules and regulations in order to maintain trade with EU countries; the UK's agriculture departments will have no say in these decisions. Whatever the outcome of the current EU pesticides review, there is increasing and widespread concern that many pesticides are used as a 'just in case' measure rather than as a last resort. This overuse raises the amount of pesticides in the environment and has implications for wildlife depletion both on land and in the soil. It may also lead to greater pesticide tolerance by pests, thus creating a potential lack of pest control.

New Furrow 10: While there is no doubt that badgers infect cattle – and that cattle infect badgers – all the objective evidence and most expert opinions point to the conclusion that culling badgers is not going to play a significant long term role in eliminating bovine TB. Continued strict and enforced biosecurity measures on farms combined with sustained and more rigorous TB testing will be the main means of eradicating the disease. Long term, badger vaccinations against TB might help control the disease.

One of the most controversial issues in farming in recent years is the culling of badgers and whether this helps control the prevalence of bovine TB in cattle in southern Britain. The farming unions and many farmers argue that culling badgers must be part of the measures implemented to bring the disease under control. Yet the one large scale, fully-monitored study of badger culling, in South West England, found that any bovine TB reduction in cattle was short lived, and that some culling exacerbated the problem by dispersing surviving badgers and increasing the incidence of the disease in cattle.

Badger culling is still a major component of TB control policy in England, though certainly not on any systematic basis that allows its efficacy to be gauged. Yet in Wales – where the Welsh Government has not yet carried out any badger culling (though is likely to on a very limited scale from 2018) – the incidence of cattle TB is falling. It is still rising in England. All of the objective evidence, and most scientific expert opinion, suggests strongly that culling badgers is not going to play a significant role in eliminating cattle TB and emphasises the need for much more tightly enforced cattle biosecurity and TB testing on farms.

Endnotes

i   'Economist calls for exit bond for farmers,' *Farmers Weekly*, 11 November 2016.

ii ' Wildlife habitat creation within the farm business,' Paul Pickford et al. Centre for Ecology and Hydrol ogy, 2016.

iii 'Wildlife-friendly farming increases crop yield: evidence for ecological intensification,' Pywell et al, 2015. Proceedings of the Royal Society B, DOI: 10.1098/rspb.2015.1740.

iv   Agricultural and forest area under environmental management schemes: B1a. Area of land in agri-envi ronment schemes. JNCC, 2017.

v   'Glastir,' Wales Audit Office. September 2014.

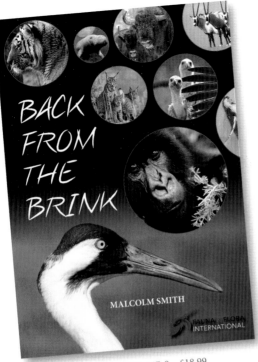

BACK FROM THE BRINK

MALCOLM SMITH

FAUNA & FLORA INTERNATIONAL

ISBN 978-184995-147-0   £18.99

Success stories of animals threatened with extinction whose fortunes have been reversed

Tales of lifetimes dedicated to wildlife conservation

Features some of the most iconic animals from around the world

An antidote to wildlife doom and gloom, this book shows that it's possible to turn the tide of species decline. It's full of conservation success stories, valuable lessons and inspiration for the future. **Wildlife World**

…the wide-ranging approach of Back from the Brink *gathers stories of personal courage, conservation efforts, special challenges, and world encounters in the effort to bring back animals from places where they had once thrived. …a solid pick for any science and nature collection.* **California Bookwatch**

…packed with information… an excellent and informative read. **BTO News**

…it's wonderful to hear any tales of animals that have been saved, and of the fantastic individuals who go to great lengths to help them. The Weekly News

Anyone with the remotest interest in conservation should read this book now. …weaves some wonderful stories into this statement of hope. … Buy it. In doing so you will contribute to that spirit. **SWARA magazine**

www.whittlespublishing.com

Gone Wild

STORIES FROM A LIFETIME OF WILDLIFE TRAVEL

MALCOLM SMITH

ISBN 978-184995-177-7   £16.99

Travel stories about some of the most amazing and unusual parts of the wild world

Vivid descriptions of an array of wild animals and the places they inhabit

Strange and amusing encounters with local people and occasional brushes with authority

...as his latest book describes, he's had scary moments, as well as awe-inspired ones! Visits to the New Forest, Iceland's offshore islands and countless wonderful corners of our planet have left Malcolm at times inspired, overawed and joyous... **The Sunday Post**

...What a fascinating book. ...describes his adventures (and misadventures) in a variety of far flung places as diverse as the Saudi Desert and the Niger River... The stories are beautifully told and demonstrate not only an incredible knowledge of wildlife world-wide but of the influence, good or bad, of humans on some species and their habitat. ... Gone Wild is a world tour for wildlife enthusiasts and is a book I would thoroughly recommend.
Wildlife Detective, the blog of Alan Stewart

www.whittlespublishing.com

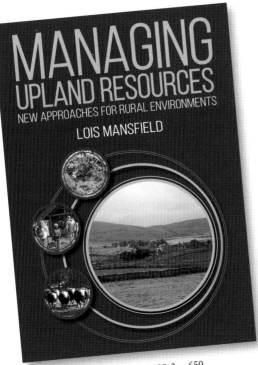

MANAGING
UPLAND RESOURCES
NEW APPROACHES FOR RURAL ENVIRONMENTS
LOIS MANSFIELD

ISBN 978-184995-229-3   £50

A comprehensive synthesis of academic and practical experience in understanding and managing upland environments

Includes models and practice guides outlining how vital areas are to be managed, supported and developed for the future

Provides practical options to create solutions

...we need more and better livelihoods in the uplands – Lois Mansfield's new book shows, with some much needed intellectual rigour, how this can be achieved. There's a nice set of tables...giving a pick-and-mix menu for upland resource management which is well worth thumbing through. **Alan Spedding, RuSource** and **Royal Agricultural Society of England Members' Agri-Bulletin**

...an invaluable blueprint for the future. ...a narrative, interspersed with diagrams, coloured charts, tables, lists of all kinds to which you can refer for evidence and information. The results of an immense amount of research are summarised and acknowledged. ...this most complete reference book for the professionals, the scientist, the academic, or the amateur upland enthusiast. It's a one-book degree-course – highly recommended. **George Macpherson, Consultant Editor, Appropriate Technology Magazine**

www.whittlespublishing.com